Christian Anti-Natalism

Kirk Lougheed

Christian Anti-Natalism

Reconsidering the Theology of Procreation

palgrave
macmillan

Kirk Lougheed
Theology Department
LCC International University
Klaipeda, Lithuania

ISBN 978-3-032-03901-9 ISBN 978-3-032-03902-6 (eBook)
https://doi.org/10.1007/978-3-032-03902-6

This Palgrave Macmillan imprint is published by the registered company Springer Nature Switzerland AG
The registered company address is: Gewerbestrasse 11, 6330 Cham, Switzerland

Acknowledgements There are many people that are owed thanks for this book. Much of this book was written during my time as director of the Center for Research on Faith and Human Flourishing at LCC International University. In this role I received significant support, especially in the form of extensive release from teaching duties. The University of Pretoria provided significant financial resources for my research, in addition to extensive use of their library resources.

As I explain in Chapter 1, it is not without significant trepidation that I decided to write this book. Now that the book is complete, I am glad I wrote it because I am more convinced than ever that the central thesis and arguments are worthy of careful consideration. There are two people that I owe thanks to for encouraging me to pursue this project. The first is Amanda Sukenick, who was enthusiastic about it from my first mention of it and pointed me towards relevant sources. Second, David Benatar encouraged me to see it through to publication. As I said, I am glad to have completed it and probably would not have done so without their encouragement.

I have the privilege of frequently interacting with many talented philosophers. I am thankful to them all for their support over the years. Special thanks to Thaddeus Metz, who has consistently gone far above and beyond what could be considered reasonable support for a mentor. Thanks also to Blake Hereth who provided extensive comments on large portions of the manuscript. I would also like to thank my editor at Palgrave Macmillan, Brendan George, who was supportive of the project from the beginning. I am also grateful to an anonymous referee for the press.

Though my family very likely continues to be puzzled by my research programme and what exactly I do, I am grateful for their constant support of my endeavors.

Toronto, Ontario
June 2025

Kirk Lougheed

Competing Interests The author has no competing interests to declare that are relevant to the content of this manuscript.

Contents

Part I

Introduction to Christian Anti-Natalism

1

The Counterintuitive Topic of Christian Anti-Natalism

1.1 Introduction

This central aim of this book is to explore the best possible defense of the thesis that if Christianity is true, then anti-natalism is true. Christians should be anti-natalists. Readers will be familiar with various versions of Christianity as it represents one of the major monotheistic religions with millions, if not billions, of adherents around the world. Anti-natalism is significantly less recognisable, representing a stance in procreative ethics which says that it is always (or almost always) impermissible to procreate. Thus, my goal is to develop a defense of the surprising and counterintuitive view that it is always (or almost always) impermissible for Christians to procreate.[1]

Ken Coates writes that:

[1] More specifically, the view I examine says that it is wrong for Christians to procreate in the actual world, though there are possible worlds where it would be permissible.

K. Lougheed, *Christian Anti-Natalism*, https://doi.org/10.1007/978-3-032-03902-6_1

> Millions of women, especially in developed countries, are choosing not to have children with apparently no instinctual urge to reproduce. Nonetheless voluntary childlessness, especially on the part of the married, remains taboo in Western 'advanced' societies. It is still considered as a form of deviant behavior (Defago 2005; Basten 2009; Overall 2012). Given the strength of conventional wisdom and the status quo it is important to pay attention to dissenting viewpoints concerning procreation. (Coates, 2014, 14)

Now, Coates is making a sociological claim that I will not defend in this book. However, inasmuch as it is safe to say that procreation is extremely common behaviour, I think he is correct that 'dissenting viewpoints' should be considered. Indeed, one of the roles of philosophy is to critically assess commonly held beliefs and practices.

It is not without some trepidation that I decided to finally write this book. The topics that I discuss in it are extremely controversial and may illicit strong negative emotional reactions from readers. Indeed, in explaining the harms associated with existence, the well-known anti-natalist philosopher David Benatar writes "[m]ost people, under the influence of powerful biological dispositions towards optimism, find this conclusion intolerable. They are still more indignant at the further implication that we should not create new people" (Benatar, 2006, 1). Of course, I worry that such negative reactions will not just be directed towards my ideas but will also be directed at *me*. Yet I ultimately decided to write this book because I believe that the thesis needs to be seriously considered by Christians. Procreative ethics is a serious matter and important truths about it should not remain unstated. Christians who are serious about living out their religious convictions authentically will have to grapple with the arguments that I present throughout the book. Whether I myself am a Christian anti-natalist has no bearing on the plausibility of the thesis of this book. For readers who are familiar with my work elsewhere, especially my other books, this methodology will come as no surprise (e.g., Lougheed, 2020a, 2022b). I believe that one effective way of philosophising is for an author to put forward arguments, defend premises, address objections, draw logical connections., etc., irrespective of what the author themselves may or may not believe. I utilize this approach here in the hope that it will help prevent distracting readers about my purported motivations for writing the book.

In the rest of this chapter, I introduce the reader to the topic in more detail, explaining that it can be understood as a project of seeking reflective equilibrium (1.2). I then turn to exploring an initial objection which says that enacting anti-natalist policies are immoral (1.3), before explaining more fully my methodological approach (1.4).

1.2 The Counterintuitive Topic of Christian Anti-Natalsim or Seeking Reflective Equilibrium

Suppose Christianity is true. Suppose also that anti-natalism is true. It is not lost on me that many readers, regardless of whether they are Christians or anti-natalists, will find this combination of views to be entirely implausible. Many will suspect, if not take it as just obvious, that Christianity and anti-natalism cannot both be true. Indeed, it might be thought that I am seeking to develop a defense of something pretty close to an outright contradiction.

From the Christian perspective, it is almost always taken as obvious that procreating is permissible, if not recommended. Though I will not defend this claim here, I suspect that many Christians think marriage, procreating, and childrearing is the best way to live one's life. A desire for children within the context of marriage does not need to be defended, while a person who remains intentionally single and childless needs to defend this choice.[2] Consider that in Genesis 1:28 God tells Adam and Eve to "Be fruitful and multiply and fill the earth." Christians are commanded by God to have children. Christians are sometimes thought to prize the family unit, including procreation and childrearing, more than many other alternative worldviews, especially secular ones. Indeed, when Jesus' disciples (showing their own ignorance of his message) ask him who is the greatest in the kingdom of heaven he points to the need

[2] Or perhaps needs to be thought of as having a special calling from God to remain single and childless as do, say, Roman Catholic Priests.

for childlike humility, saying "[w]hoever receives one such child in my name receives me, but whoever causes one of these little ones who believe in me to sin, it would be better for him to have a great millstone fastened around his neck and to drowned in the depths of the sea" (Matthew 18: 5–6). God values children. Finally, Christians are taught to be thankful to God for all the good things in life, including for their very existence. Life is a precious gift from God. The classical theistic tradition affirms that existence (i.e., essence) is intrinsically good. It appears that anti-natalism is a nonstarter for the Christian.

From the perspective of anti-natalism, almost all of the contemporary philosophical literature appears to assume that metaphysical naturalism is true. Though little consideration has been given to the plausibility of anti-natalism in the context of different religious traditions, toward the end of his influential book, *Better Never to Have Been: The Harm of Coming into Existence* (2006), Benatar argues that even in the Christian tradition there are exceptions to the 'fruitful' verses in the Bible. Catholicism exempts the clergy, and procreation by non-sexual means.[3] Shakers believes that everyone should be celibate (Benatar, 2006, 221). He also observes that:

> [B]oth Jeremiah and Job ru[e] their births. Job regrets his having been conceived and the fact that he did not die *in utero* or at birth. Jeremiah goes further and curses the man who did not abort him. It is striking how different such views are from those of the cheery fundamentalist with an unsophisticated, monolithic view of the right way. Whereas Jeremiah and Job think and speak freely—even challenge God himself—all too few religious believers follow suit. For them piety precludes such critical thinking and speaking (Benatar, 2006, 222)

Benatar's point is that religious thinking on such matters cannot be taken as a monolith. Indeed, Benatar also observes a debate about whether or not it was better that humans had been created between the early rabbinic schools of the House of Hillel and the House of Shammai. Indeed, he explains that "[t]he Talmud relates that these two schools

[3] My Lougheed (2023) piece is another exception.

debated the matter for two and a half years and the issue was eventually settled in favour of the House of Shammai [i.e., on the side that it is better humans were not created]. This is particularly noteworthy, because in cases of disagreement between these two schools, the law almost always follows the House of Hillel" (Benatar, 2006, 223). Having said this, it still seems safe to say that the default position of monotheists in general, and Christians in particular, is that it is better to have been born and procreation is good. The religious exceptions and precedents of anti-natalism pointed to by Benatar have hardly been widely adopted. Likewise, Benatar does not spend significant time defending the fact that anti-natalism is consistent with various religious traditions, and doing so is clearly not his primary aim. The Biblical and Historical record for anti-natalism will be investigated further in the next chapter.

One potentially helpful way of understanding the purpose of this book is that I am seeking a way to establish reflective equilibrium for the person who finds themselves believing the truth of both Christianity and anti-natalism. How can these two views be coherently held by someone at the same time? I show that this might be possible by developing an argument which focuses not just on the exposure to the risk of certain harms in one's earthly life, but to the exposure of the risk of hell in the afterlife. I suggest that some widely accepted commonsense intuitions about the level of risk it is permissible to expose a non-consenting adult to entail that Christians should be anti-natalists. Reflective equilibrium is established by demonstrating that not only is Christianity consistent with anti-natalism but that it actually recommends it. Again, I do not defend this position definitively, but instead I am attempting to construct the best possible defense of Christian anti-natalism. Thinking of this in terms of attempting to establish reflective equilibrium between the two positions is helpful, and as shown from Benatar's remarks above and the discussion in the following chapter, the inconsistency of Christianity and anti-natalism is probably too quickly assumed.

1.3 Is Christian Anti-Natalism Beyond the Pale?

Anti-natalism generates visceral reactions. For example, the work of Benatar, particularly his book, *Better Never to Have Been: The Harm of Coming into Existence*, has been subject to a number of emotionally charged and unfair criticisms where the conclusion of anti-natalism, not his specific arguments, is clearly the target. Surprisingly, some such reviewers even admit to not even having read the book. Being aware of its conclusion is apparently all that is needed for some to criticize a work. What is further surprising is that these pieces have sometimes appeared in peer reviewed philosophical venues where the focus should be on the philosophical merit of the arguments in question. Though Benatar defends anti-natalism as applying to everyone, I take it that arguing Christians ought to be anti-natalism is no less controversial. It is therefore no less likely that its conclusion may similarly upset certain people.

In his article, "Ethical Unthinkabilities and Philosophical Seriousness," Sami Pihlström attempts to articulate the problem with discussing, let alone putting forward a defense of anti-natalism (2009).[4] He writes that there may be "ideas that are dangerous enough not to deserve serious argumentative attention. Perhaps there are philosophical ideas and arguments that ought to be left aside precisely because they violate some human values and ideals that are cherished more deeply than the ideal of sound argument itself" (Pihlström, 2009, 657). However, Pihlström does not offer any substantive philosophical objections to Benatar's arguments. He explains that his "goal is not simply to dispute Benatar's view, which is in my view too far off the humanly possible moral scale even to be seriously disagreed with. More precisely, I cannot find myself disagreeing with Benatar, because I would have to share a sufficient common ground with him in order to be able either to agree or

[4] Nicholas Smyth has framed a similar objection in more charitable terms, arguing that discussions of anti-natalism are too impersonal in nature. However, though he uses anti-natalism as an example, his criticisms are intended to imply to procreative ethics more broadly. He suggests that discussions in applied ethics need to be existentially grounded (2020).

to disagree" (Pihlström, 2009, 657). Pihlström denies that 'argumentation' is the most important thing in 'moral philosophy' and that there are certain ideas that ought to be simply ignored because they pose an 'ethical unthinkability' (Pihlström, 2009, 657). There is a "moral demand that truly serious philosophizing should not only be argumentatively and intellectually skillful but also be a humanly, or even existentially, *honest* activity" (Pihlström, 2009, 658). The claim here does not seem to be about truth-telling but honesty in the sense of defending a conclusion that could be accepted in the first place.

But specifically *why* does Pihlström believe anti-natalism is beyond the pale? It is not always entirely clear. One possible candidate I can glean is that procreation and the perpetuation of the human species is such a fundamental human activity that it is simply not a candidate that is open to philosophical questioning (Pihlström, 2009, 662). This does not mean that any argument for anti-natalism is unsound, but that the view "is so remote from truly humanly considerable, humanly 'thinkable,' thought structures that he [i.e., Benatar] writes himself out of the game" (Pihlström, 2009, 662). Pihlström does not believe speech or expression should be restricted. Instead, he suggests "that moral philosophers should be extremely cautious even in starting to follow—to read and think about—arguments like Benatar's. What may initially seem a more or less plausible premise will turn out to lead to dangerous, ethically monstrous conclusions" (Pihlström, 2009, 668). So, though Pihlström does not believe Benatar's book should be burned (or presumably this one), he does suggest that people should not read it. Again, though Pihlström's target example of an 'ethical unthinkability' is Benatar's anti-natalism, I see little reason to think Chrisitan anti-natalism is any less controversial such that his concerns would not also apply here.

One problem with Pihlström's view is that it may well be self-referentially incoherent or at least fail to take its own advice. Pihlström is platforming and discussing one of the very views he argues that philosophers ought to ignore (Benatar, 2011, 605–606). Pihlström's article may also be self-defeating in that it draws *additional* attention to anti-natalism (Benatar, 2011, 606).[5] He also gives *arguments* for the conclusion that arguments are not the best or most conclusive way of philosophizing.

Pihlström's claim that anti-natalism should not be considered is at least partly the result of his very different perspective about the goals of philosophy. In Benatar's response to Pihlström, he explains that:

> Philosophers aim to take the owl's-eye view. Just as the owl sees even in darkness, so philosophers aim to evaluate views even when they are obscured in the dim light of convention and orthodoxy. It is thus unsurprising that the owl (of Athena or Minerva) is the symbol of philosophy. Professor Pihlström has an alternative avian hero. Like the proverbial ostrich that buries its head in the sand, he would have us ignore rather than engage purportedly unthinkable views. This, he thinks, is the philosophically serious approach. (2011, 606)

Part of the problem with Pihlström's approach is that it is too easy to dismiss a view one does not like by simply claiming it is beyond the pale, and not a genuine option (Benatar, 2011, 607). Who decides which views are genuine options? What if I find Pihlström's claim that certain philosophical ideas should not be discussed to be the one idea that is beyond the pale? A society not structured on race is difficult to imagine for the racist, just as a society with gender equality cannot be imagined by the sexist (Benatar, 2011, 607). This does not mean they are not genuine options, but merely that these people have rather limited imaginations.

If Christian anti-natalism really is a dangerous idea, then it should be faced head on and refuted by careful argumentation.[6] Likewise, philosophy is not merely about constructing and refuting arguments. With Benatar, I think "that it must be a serious attempt to reflect on our lives" (Benatar, 2011, 614). And "I do not think that our reflections should be hidebound by convention and the horizons of pedestrian thought. The job of the philosopher is often to question commonly held assumptions.

---

[5] Benatar only briefly acknowledges these issues of consistency in Pihlström and does not focus his reply on them (2011).

[6] Benatar also notes that if an idea really is dangerous, it is all the more important to carefully refute it. Likewise, he maintains that since anti-natalism ultimately reduces the amount of harm in the world, it is difficult to see how his ideas could be dangerous (Benatar 2006, 612–613).

It is the job of other philosophers to consider those challenges seriously rather than to bury their heads in the sand" (Benatar, 2011, 614).[7]

1.4 The Policy Objection

A slightly different worry from the one just raised in 1.3 that could be used to forestall any further inquiry into anti-natalism says that adopting anti-natalist policies would involve widespread human rights abuses. Indeed, in arguing that the world's population needs to decrease because of the climate crisis, Trevor Hedberg observes that population size is almost never mentioned by governments or organizations who claim to care deeply about the planet (2020). He explains that "[e]ven among the environmentally conscientious, discussion of population are sometimes avoided because of their rhetorical challenges and worries about the fraught, morally unsettling history of previous population policies" (Hedberg, 2020, 3). Though population growth is having a very real and negative impact on the climate, from the 1980s onwards even climate scholars said very little about it. This has only recently started to change in scholarly discussions (Hedberg, 2020, 4).[8] Hedberg speculates (and rightly, I think) that part of the general reluctance to discuss population is a strong desire not to be associated with problematic policies. He explains that for a time:

> [D]iscussion of rising population and the connection between population growth and environmental destruction became politically incorrect because suggestions to stabilize or reduce population were perceived as

[7] Consider: "There is a common theme running through some of Professor Overall's arguments, namely that the anti-natalist conclusion is deeply counter-intuitive and should be rejected (at least in part) for that reason. The counter-intuitiveness (to most people) of anti-natalism has never been lost on me. The question is how much store we should put on that widespread sense of counter-intuitiveness, given that procreation leads to so much suffering and death. After all, any view that permits the imposition of a massive risk of severe suffering and the certainty of death is – or should be – at least as counter-intuitive, if not more so." (Benatar, 2022, 132).

[8] Hedberg cites a number of recent examples of such discussions: Clarke and Haraway (2018), Conly (2016), Coole (2018), Crist (2019), and Rieder (2016).

> disadvantageous to women. Recent history seemed to validate this perception: measures such as China's one-child policy and coercive episodes of family planning in India appear morally unacceptable. (Hedberg, 2020, 5)

Though Hedberg is writing in the context of population growth in connection to the climate crisis, I believe that such sentiments apply equally, if not more so, to discussions of anti-natalism.

In the beginning of *Better Never to Have Been: The Harm of Coming into Existence*, Benatar writes that:

> Given the deep resistance to the views I shall be defending, I have no expectation that this book or its arguments will have any impact on baby-making. Procreation will continue undeterred, causing a vast amount of harm. I have written this book, then, not under the illusion that it will make (much) difference to the number of people there will be but rather from the opinion that what I have to say needs to be said whether or not it is accepted (2006, vii).

I share Benatar's sentiments about the impact of his book. The academic work of a handful of professional philosophers is not going to help change or create any anti-natalist policies. Indeed, though it appears Benatar's work has reached people beyond the academy, it is still only a miniscule percentage of the global population who have read his work, let alone who agree with him. Not only is Benatar right that anti-natalism will never be widely adopted given the natural drive to procreate, not to mention other extremely powerful social conventions, but developing anti-natalist policies is not the goal of his work. *Nor is it mine.* I am merely aiming to explore the best possible case for the claim that if Christianity is true, then anti-natalism is true. I am making a theoretical point; I am *not* claiming that Christian communities ought to develop and enforce anti-natalist policies. The soundness of the theoretical arguments I develop in this book are entirely independent of whether it is possible either practically or morally for Christian communities to enact anti-natalist policies. Christianity could entail anti-natalism even if it is immoral to enforce anti-natalist policies.[9] I am therefore concerned with what people should *believe* about the truth value of anti-natalism given Christianity, not whether they could or should enact policies to enforce it.[10]

Finally, debates about policy often involve empirical claims and so in this case is really a subject for sociologists. Consider that in objecting to the anti-natalist arguments from Benatar, Christine Overall writes that:

> Benatar is surprisingly oblivious to the implications of his theory for women's rights and well-being. Most of his discussion of procreation is curiously gender neutral. At the same time, his theory implies that women's reproductive labor produces bad consequences. That is, the idea that it is better in every case never to have been implies that women's reproductive labor in pregnancy, birth, breastfeeding, and even rearing children contributes to the accumulation of net harm on this planet. (Overall, 2012, 115)

Overall continues to insist that women would not be better off if anti-natalist policies were adopted, especially in societies where they are primarily valued for their ability to procreate. She speculates that "[i]f Benatar's theory were to gain credence (unlikely though that may be), then one of women's primary social contributions, recognized even (or especially) in the most misogynist societies, would be seen as a liability. Might this view lead to an increase in the infanticide of girls or to assaults on pregnant women?" (Overall, 2012, 115). Benatar has responded that adopting anti-natalism would have just the opposite effect. It would liberate women. It would be impossible to see them primarily as child bearers and rearers (Benatar, 2019). This discussion highlights my point: it is simply an empirical question whether anti-natalist policies would help or harm women. And I am not conducting a study about the effects of adopting anti-natalism.

---

9 This does not violate 'ought implies can'. Everyone *could* cease to procreate of their own accord without any coercive policies in place.

10 Benatar also notes that because there is reasonable disagreement about anti-natalism, it would be wrong to enact a policy enforcing it.

1.5 Methodology and the Title of This Book

Prior to concluding this chapter, it is informative to reflect on the methodology I will employ in this book. The subtitle of this book is *Reconsidering the Theology of Procreation.* However, this book is primarily *not* a work in theology or theological ethics. Though later I will obviously have something to say about Christianity in order to offer the argument, my focus is *not* theological. In light of this, the reader may rightly wonder why the word theology appears in the title at all.[11] 'Christian ethic' or 'philosophical theology' are both probably more accurate, but do not work well in the title. Finally, inasmuch as a Christian considers the permissibility (or more strongly, the requirement) to procreate as resulting from their Christian beliefs, it is reasonable to classify the belief as at least partly theological. However, this book should not be mistaken for the work of a theologian. Finally, replacing 'theology' with 'ethics' would also not work well because it might imply an ethical argument that applies to everyone instead of just Christians. It risks creating confusion about the scope of the work.

I am a philosopher and will use the tools of contemporary 'analytic' or Anglo-American philosophy including, but not limited to, conceptual analysis, attention to definitions and assumptions, and rigorous argumentation. The reader looking for extensive Biblical exegesis, church history, systematic theology, etc., will not find it in this book. Having said that, I hope that the book will be of interest to theologians and scholars of religion across different disciplines who have an interest in Christianity and procreative ethics. Indeed, I hope such scholars will engage with the arguments in this book using the tools of their own disciplines.

[11] Indeed, the idea of the subtitle comes from my editor at Palgrave, Brendan George.

1.6 Conclusion

I will not belabour the introduction to this work with further clarification or hedging. To reiterate, I am developing arguments for the thesis that Christians ought to be anti-natalists, while remaining silent on my beliefs in order to keep the focus on the substance of the arguments. I recognise that the default position is that Christianity and anti-natalism are incompatible with each other. One way of understanding this book, then, is to conceive of it as attempting to achieve reflective equilibrium in a person who is both a Christian and anti-natalist. Finally, nothing in this book should be taken as advocating for anti-natalist policies. In the next chapter, I offer a brief overview of anti-natalist ideas as primarily found in the Christian tradition including in the Bible and Church Fathers in order to better situate the main arguments to follow.

2

Anti-Natalism and Christianity: Historical, Social, and Biblical Precedents

2.1 Introduction

It would be difficult to fault someone for thinking that anti-natalism officially emerged with David Benatar's book, *Better Never to Have Been: The Harm of Coming into Existence* (2006). Indeed, the contemporary Anglo-American philosophical literature often implicitly takes Benatar's book as its starting point, even when there are handwaving acknowledgements to earlier statements.[1] This perhaps might make it feel especially ad hoc to try to show that Christianity, an ancient world religion, is compatible with and may even entail the truth of anti-natalism.

It might therefore be surprising for readers to learn that some scholars do not believe the Bible unreservedly teaches pro-natalism. Likewise, some believe that anti-natalist thinking in various forms has existed throughout much of the Christian tradition. My purpose in this chapter is *not* to suggest that because anti-natalist thinking has existed

[1] In fact, though his book is usually the starting place, Benatar defends anti-natalist views in two earlier journal articles (1997, 2001), in addition to discussing related themes in other earlier work (1999, 2000).

K. Lougheed, *Christian Anti-Natalism*, https://doi.org/10.1007/978-3-032-03902-6_2

throughout the history of Christianity that therefore Christians should be anti-natalists. Instead, I hope to make the idea of Christian anti-natalism more palatable, if only ever so slightly, to Christians who are likely to balk at the suggestion that ant-natalism could ever be consistent with their religio-ethical commitments. In short, I attempt to show that there is at least some precedent for anti-natalism in the Christian tradition.

I begin by examining the well-known and oft-cited pro-natalist 'fruitful' verses in the Bible (2.2) before turning to lesser discussed anti-natalist texts in the Bible (2.3). After this, I examine some historically influential Christian thinkers, showing the context in which their pro-natalist thinking emerges (2.4). I then turn to anti-natalist precedents that appear outside of the Christian tradition (2.5). It turns out that not only is anti-natalist thinking probably not new within the Christian tradition, it is hardly new generally. There are precedents for it throughout much of human history. Finally, this chapter is intentionally quote heavy as I am trying to let the authors speak for themselves.

2.2 Pro-Natalism in the Bible

John Mckeown's wide-ranging book, *God's Babies: Natalism and Bible Interpretation in Modern America,* examines a whole host of pro-natalist historical precedents in the Christian tradition including from Luther, Augustine, and the verses of the Bible often appealed to as pro-natalist. Though writing in the context of studying American Christianity, much of the work is still relevant in exploring the biblical and historical precedents surrounding procreation in the Christian tradition.

According to Mckeown, the most cited Bible verses in favour of pro-natalism are Gensis 1:28 (ESV):

> And God blessed them. And God said to them, "Be fruitful and multiply and fill the earth and subdue it, and have dominion over the fish of the sea and over the birds of the heavens and over every living thing that moves on the earth.

And Psalm 127: 3–5 (ESV):

> Behold, children are a heritage from the Lord, the fruit of the womb a reward. Like arrows in the hand of a warrior are the children of one's youth. Blessed is the man who fills his quiver with them! He shall not be put to shame when he speaks with his enemies in the gate.[2]

These verses are often appealed to in defense of pro-natalism, even if their specific scope is debated (Mckeown, 2014, 17). For example, Mckeown draws a distinction between different types of pro-natalists (i.e., those who hold different views about contraception and the purposes of sexual acts). However, these distinctions are not important for my purposes.[3] I am simply aiming to show that there are plausible interpretations of these verses that are not pro-natalist. At the very least, I show that pro-natalist implications cannot be assumed of them without argument.[4]

Different interpretations of the fruitful verses include that "they can be construed as cultural pragmatism, or as an accommodation to a prosperity cult, or as ethnocentric and competitive xenophobia, depending on differing views about the emphases of Old Testament theology" (Mckeown, 2014, 26). Mckeown also claims that even if the original meaning of an Old Testament verse were agreed upon, it is doubtful this would be "a suitable guide for Christian behaviour" (2014, 26). Mckeown makes two primary points about the fruitful verses in the Old Testament. First, they are related to prosperity. Second, contemporary pro-natalists fail to address certain contextual features, including the fact that there was an extremely strong preference for sons (Mckeown, 2014, 109).

Consider first the demographic context of the Ancient Near East. There were extremely high premature mortality rates, and more than

---

[2] Mckeown also points to several other verses and his names them collectively 'fruitful verses' (see 2014, 215-222).

[3] For more on this see Doriani (1993), Mangina (2004), Poulson (2006) and Provan (1989).

[4] Mckeown observes that the pro-natalists implications of these verses has even been challenges in *Christianity Today*, albeit mostly responding to critiques of contraception (2014, 20; see Van Leeuwen, 2001).

a third died before the age of five (Mckeown, 2014, 111). The vast majority of Israelites were farmers or herders:

> The writers of the Old Testament were scribes and priests. Though they were part of a religious elite with wider concerns than farming, around 95% of ancient Israelites were farmers (Blenkinsopp [1997], 54), and the writers would have been aware of agricultural concerns. It is helpful to consider the contribution of material culture to Old Testament ideas about fertility. In brief, ancient farmers would normally esteem prolific human fecundity, and that was accentuated in early Israel due to their circumstances. (Mckeown, 2014, 112)

The first chapter of Genesis may at least partly be a response to Ancient Near East fertility cults:

> Presenting fertility as a blessing from the one creator may also be polemic against ancient Near Eastern fertility cults. Leading archaeologist William Dever emphasizes the "central role of sex and reproduction in Canaanite religion" (Dever, "Who" 199). John Hartley suggests that most people in the ancient Near East "believed that fertility rites practiced at local shrines enabled their lands, flocks and wives to produce abundantly" (49). Israelite popular folk religion was perhaps similar. Figurines of a pregnant or lactating woman have been found all over Judah from the time of the Davidic dynasty, perhaps linked to a cult of Asherah and probably connected to prayers for fertility. Victor Hamilton observes that fertility rituals were often associated with the retelling of creation stories (139). Westermann suggests that Genesis 1:28 is designed to warn its hearers that when they seek fertility (the words of the blessing may derive from a traditional marriage blessing), they should not seek help from other gods because fertility is a gift from the God of Israel, and since God gave all life the capacity to reproduce at its origin, no subsequent ritual is needed (*Genesis* 161). Genesis chapter 1 is thus "a deliberate rejection of the fertility cult (Cohen 44). (Mckeown, 2014, 124)

Some commentators have suggested that the book of Genesis is written in response to the eighteenth century BC Babylonian epic *Altrahasis*, Mckeown explains that "[i]n that writing, the lesser gods complain about their work of farming and maintaining canals, so humans are created as laborers. Men, however, multiply and become too noisy, their "uproar" disturbing the peace of the gods" (127–128). Consider the additional scholarship that Mckeown summarises:

> William Moran identifies the words "be fruitful and multiply" after the Flood (Genesis 9:1) as a "conscious rejection" of *Atrahasis*' presentation of the "limitation of man's growth" as a remedy for disorder (61). Anne Kilmer similarly argues that whereas *Atrahasis* calls for man to "limit his increase ... the biblical text indicates the opposite command" (174). Isaac Kikawada argues that here "God, far from punishing man for population growth, is rather ordering him 'be fruitful and multiply and fill the earth' ... this command ... was argumentative, almost polemical, in its original context" (38–39). This is speculative, assuming that *Atrahasis* was known to the biblical writer, but plausible given Israelite natalism. (Mckeown, 2014, 128)

With respect to the scholarship on Psalm 127: 3–5, Mckeown writes that:

> Many scholars consider that this Psalm was developed from a pre-existing proverb, and identify the genre of its origins as proverbial wisdom for ordinary people about "everyday life" (Kraus 453) and "the farmhouse" (Hunter 237). It was "domestic" (Gerstenberger 346), and spoke to the "universal preoccupations" of work, security, and family (Kidner 440). The notion that "the larger the family, the less vulnerable" reflects ancient culture (Mays 401). However, most scholars believe that, whatever its origins, Psalm 127 was reshaped to convey theological messages. (Mckeown, 2014, 130)

One consistent theme in Mckeown's book is the claim that the fruitful verses in the Old Testament are addressed to the ancient Israelites and their particular context, which means that it is illegitimate for Christians to infer universal pro-natalist support from them (2014, 133). In sum, these ancient 'Israel-oriented' promises cannot be applied by the Christian tradition to everyone. Historically, when such promises were taken to apply to the church, their meaning was changed to hope for spiritual fulfilment instead of children. However, "[n]atalists have pulled apart the Christian tradition's approach to these verses, happily adopting its transference of addressee, but neglecting the transformation of meaning (Mckeown, 2014, 133).

In the Old Testament, the benefit of having children was for protection and labourers. Fertility blessings were therefore about prospering and flourishing. Notice that this means that if children did not in fact bring prosperity, then they would fail to be a blessing and it would be difficult to see how Old Testament blessings could apply to them (Mckeown, 2014, 134). In other words, 'be fruitful and increase' "is qualified and limited by being a blessing. [This means that] if [the] population rises to a point at which it is detrimental to the flourishing of some human individuals or nations then it is certainly not a blessing" (Mckeown, 2014, 134). Mckeown's point is that the command to procreation is contingent because the blessings associated with it are contingent. In sum:

> [T]he cultural, economic, political, and theological context of the Old Testament fruitful verses limits their plausible appropriation by modern natalism. Pro-fertility ideas in the ancient Near East are unsurprising given the demands of immortality through patriarchal lineage, agricultural subsistence in the Judean highlands, and a small people struggling to retain their national identity against hostile empires. (Mckeown, 2014, 143)

2.3 Anti-Natalism in the Bible

Observe that there are numerous places throughout the Bible that express anti-natalist sentiments.[5] Let's start by considering some passages from the Old Testament:

> I saw all the oppressions that are done under the sun. And behold, the tears of the oppressed, and they had no one to comfort them! [...] And I thought the dead who are already dead more fortunate than the living

[5] Many of these citations are from de Giraud (2021).

who are still alive. But better than both is he who has not yet been has not seen the evil deeds that are done under the sun. (Ecclesiastes 4:1–3)

A good name is better than precious ointment, and the day of death than the day of birth (Ecclesiastes 7:1).

After the destruction of Jerusalem, the prophet Jeremiah remained single and childless while appearing to recommend that it's better not to reproduce in tragic circumstances (Jeremiah 16:1–4). He later expresses disgust at having been born (Jeremiah 20:14–18).

Likewise, fertility is not always portrayed unambiguously as a blessing in the Old Testament:

For six generations which span hundreds of years, Cain's lineage prospers genetically with the multi-generational fecundity that [...] modern natalist would-be patriarchs dream of and plan. But their fecundity is not a sign of God's favor: Cain is especially cursed (4:11), and his great lineage is doomed as generations later every one of its offspring is destroyed by God's Flood, exposing all their mothers' labors as futile". (de Giraud, 2021, 125)

All except for one line from Seth in Noah's family are destroyed in the flood nine generations later.

Théophile de Giraud suggests that Jesus is unconcerned with the biological family. While 'Honour your mother and father' is one of the ten commandments, "Jesus demonstrates an astonishing contempt for the elementary signs of respect owed to parents. Even in his childhood, concomitantly with the episode of the debate in the Temple with the doctors of the law, one discovers a 'runaway' Jesus deserting parental authority and completely indifferent to the anguish his long absence inflicts on Joseph and Mary" (de Giraud, 2021 13–14; see also Luke 2:41–50). His recruitment of the disciples almost always involves having them disregard their biological families (Matthew 8: 21–23; Luke 9:59–60). According to de Giraud, it is clear that:

Jesus is in no way the apostle of the biological family. The family he promotes is of the spiritual type and his preaching clearly aims to break

> the bonds of flesh and blood in favour of those of the Spirit. The very essence of the coming of Christ is indeed to redeem Adam's sin, which has been passed down from parents to children since the Fall and the expulsion from Eden. (de Giraud, 2021, 14–15)

The best way to avoid the perpetuation of evil and original sin is simply to stop reproducing (de Giraud, 2021, 15). Jesus shows an immense love for children *who already exist* which is best interpreted as love and respect for the weakest members of society (Matthew 18:10; 7:27). He heals children (Matthew 9:18–25; 17; 14–18; John 4:46–53), blesses children (Mark 10:16), and commands his followers to receive them as if they were receiving himself (Matthew 18:5; Mark 9:37). As will be explained in below, the best way to avoid harming children is simply not to have them at all.

The Christianity that emerges in the New Testament "exalts, it is the virtues of spiritual 'fecundity', namely: the ethics of love, altruism and forgiveness, ethics through which are transcended, rightly, our most condemnable instinctual impulses" (de Giraud, 2021, 19). This is a departure from traditional Judaism which strongly encouraged marriage, even for its prophets (with Jeremiah and Elijah appearing as exceptions in their singleness) (de Giraud, 2021, 19).

In the New Testament we find Jesus claiming that:

> Not everyone can receive this saying, but only those to whom it is given. For there are eunuchs who have been so from birth, and there are eunuchs who have been made eunuchs by men, and there are eunuchs who have made themselves eunuchs for the sake of the kingdom of heaven. Let the one who is able to receive this receive it. (Matthew 19:11–12)

> The sons of this age marry and are given in marriage, but those who are considered worthy to attain that age and to the resurrection from the dead neither marry nor are given in marriage. (Luke 20: 34–35)

Apart from the gospel narratives, we find the Apostle Paul begrudgingly acknowledging that though singleness is preferable to marriage, it is better to marry than commit sexual sins (1 Corinthians 7: 1–9).

However, he consistently devalues bodily pleasures (see Romans and Galatians), even saying of those who do marry, "Let those who have wives live as though they had none" (1 Corinthians 7:29). He writes that it is better to remain a virgin (1 Corinthians 7:25–26) and wishes that everyone adopted his celibate and childless lifestyle (1 Corinthians 7:7). Finally, the sheer quantity of reproduction cannot be seen as intrinsically good. Just consider the plagues in Egypt where the reproduction of frogs is thought terrible (Mckeown, 2014, 194).[6]

2.4 Influential Figures in the Church

Mckeown argues that though some scholars (e.g., see Bratton, 1992; Fagley, 1960) "accept the stereotype of Christian asceticism as a two-tier system whose ideal was a small celibate elite shepherding a flock of married people who were expected to breed prolifically. That picture may have been true of 19th-century European nationalist versions of Catholicism, but it has been unfairly projected back on to Augustine and the patristic period" (2014, 27).

Early Church Fathers

Mckeown argues that the Church Fathers thought the command to procreate was not unconditional given Jesus was childless. Jesus did not marry and have kids and so doing so must be seen as an optional gift.[7] Since Christ, procreation is not divinely commanded. Mckeown adds that:

> Surveys of early Christian writing (in East and West) find a consensus that the command to "be fruitful" was temporary and had been abolished (243, 37). For example, Tertullian says the new covenant "abolished

[6] Mckeown has observes that contemporary theological writing on the environmental crisis almost never mentions population (Mckeown, 2014, 110; see Keller, 1995). Indeed, worries about eugenics or a link between family planning and abortion make this unsurprising (Mckeown, 2014, 182).

[7] Mckeown observes this is Barth's view (2014, 66–67; see Barth, 1999, 268).

> the ancient command to increase and multiply" (ANF 4.40). Cyprian, a leading African bishop born around AD 200, observes that "the first decree commanded to increase and to multiply; the second enjoined continence" (ANF 5.436). Basil the Great wrote "to every one who is thinking about marriage I testify that, 'the fashion of this world passeth away'... If he improperly quotes the charge 'Increase and multiply,' I laugh at him, for not discerning the signs of the times" (NPNF2 8:214). Jerome cites 1 Corinthians 7:29 and explains that "in accordance with the difference in time and circumstance one rule applied to the former, another to us (NPNF2 6:344). (Mckeown, 2014, 67)

And:

> In the early church many married couples had children in the early years of their marriages and then entered a state of marital continence. A few went further, only reproducing minimally as a concession to ancestral and social expectations, and they are praised by patristic writers. For example, Melania agreed to bear one son and then, aged twenty, she and Pinianus settled into continence. Similarly, Therasia and Paulinus of Nola limited themselves to one son (Brown 409). Continuous childbearing by married women is not normative for Christians. (Mckeown, 2014, 70)

Though some pro-natalists recognise the high rate of singleness among early church fathers, "This anomaly is dealt with by treating singleness as an exceptional state requiring a special gift and individual calling from God. A few natalists argue that commendation of singleness in 1st century Christianity was rooted in a belief that the end of the world was imminent" (Mckeown, 2014, 71). Consider also that Athenagoras, a second century Chrisitan thinker writes that "you would find many among us, both men and women, growing old unmarried, in the hope of living in closer communion with God" (ANF 2.147 quoted in Mckeown, 2014, 71). Larry Yarbrough suggests that "Paul's silence about children and the benefits of married life was due not simply to the imminence of the end of the age, but also to the inappropriateness of most of the common arguments in favor of being married and producing children" (Yarbrough, 1986, 108 quoted in Mckeown, 2014, 71).

After the New Testament, "the anti-natalist tendencies of the early Church Fathers is rather striking, culminating in canon 33 of Synod of Elvira (303) which formally forbids clergy from having wives and children" (Giraud, 2021, 20). Indeed, "Origen, one of the leading fathers of biblical exegesis, will take this invitation very seriously and will have his testicles cut off, which mightily demonstrates that the early Christian theologians had very well understood that this new religion had come to 'block our species', according to Kierkegaard's formula" (de Giraud, 2021, 20)[8] While this list is far from exhaustive consider that Tertullian, Origen, Cyprian of Carthage, Methodius of Olympus, Athanasius, Gregory Nazianzus, Basil of Caesarea, Gregory of Nyssa, Ambrose, Chrysostom, Jerome, and Augustine all embraced remaining childless (de Giraud, 2021, 26). Consider what some of the church fathers writes:

> The bodily procreation of children [...] is more an embarking upon death than upon life for man. Corruption has its beginning in birth and those who refrain from procreation through virginity themselves bring about a cancellation of death by preventing it from advancing further because of them. (Gregory of Nyssa quoted in de Giraud, 2021, 26)

> Marriage, this mortal and servile garment [...] is the consequence of disobedience, of curse, of death. Where is death, there is marriage; remove one, the other disappears.
> The true generous ones, the lovers of light [...] fly towards the heights and get close to the Heavens, having abandoned everything here below, marriage, wealth, worries and everything which, ordinarily, attracts us to the earth. However do not believe that this permission to marry, granted at the beginning [of human history], is for the rest of time an obligation which prevents us from abstaining from marriage. For [God] wants us to renounce it.
> You are still talking to me about flocks of sheep and herds of oxen, of marriage and children? [...] Virginity is far superior to marriage. (John Chrysostom quoted on de Giraud, 2021, 27)

[8] See Brown for more on anti-natalism and virginity in the early church fathers.

> It would be utterly foolish to undergo this burdensome tribulation of the flesh, which the Apostle presages for those about to marry, by indulging in marriage in this day and age, when no service is done to Christ's future coming by begetting offspring for him through the progeny of the flesh. (Augustine quoted in de Giraud, 2021, 27)

The little known ascetic Hieracas of Leontopolis taught that marriage was forbidden to Christians. The Old Law permitted it while the New Law established by Christ forbids it.[9] The Gnostics (who considered themselves Christians) eschewed procreation.[10] Clement of Alexandria (150–215) adds:

Not wishing to help replenish the world made by the demiurge, the Marcionites decreed abstention from matrimony, defying their creator and hastening to the Good One who has called them and who, they say, is God in a difference sense: wherefore, wishing to leave nothing of their own here, they turn abstemious not from a moral principle but from hostility to their maker and unwillingness to use his creation (Clement of Alexandria, 1963, 144n quoted in Akerma, 2020, 137).

Other branches in early Christianity (or at least groups who considered themselves Christians) rejected procreation including Manichaeism, Priscillianism, the Bogomils, the Cathars (de Giraud, 2021, 31).

Finally, in a rather striking and provocative historical claim, de Giraud argues that permitting marriage and procreation was actually a survival strategy on the part of early Christians. The Roman emperor Augustus passed strict laws against celibacy around the time of Jesus' birth. For "every self-respecting empire needs soldiers, craftsmen, peasants and fertile women to survive" (de Giraud, 2021, 32). Consider that "Rome's persecutions against Christians, as early as the first centuries, are notorious. Probably tired of running to martyrdom, the Church could only engage in a double game: *better to celebrate marriages than to burn*, to paraphrase Paul. Opportunism and diplomacy which would become fatal to the evangelic ideal of non-reproduction" (de Giraud, 2021, 32). On this view, pro-natalism is a survival strategy of the early church.

---

[9] See Roldanus (1977, 341).

[10] See also Jonas (2001).

Now, de Giraud admittedly writes in a hyperbolic style, often taking his highly controversial interpretations of early Christianity as obvious. Proof-texting is a dangerous game given that it is suspiciously easy to find the conclusion one seeks in the Biblical text. My reason for engaging in it briefly here is *not* to argue for Christian anti-natalism by way of proof-texting, but instead to emphasize that anti-natalism is not necessarily an ad hoc addition to the Christian tradition. Much more serious historical work would have to be done to definitively demonstrate the conclusions de Giraud asserts with such confidence. My point in appealing to his work is more modest in showing that pro-natalism cannot simply be assumed as the default Christian position.[11]

Martin Luther

Mckeown also argues that Martin Luther's influence drastically changed attitudes about marriage and procreation from the sixteenth century onwards, not only in the protestant tradition but the Catholic one too. Consider that:

> Before him [i.e., Luther] the primary models of ideal Christian leadership had been celibate Jesus with his twelve disciples, or celibate Paul with his missionary associates and the planted churches, or later a bishop with a cathedral fellowship of celibate canons. After Luther the model was a homely pastor with a large family: a model to be imitated by the congregation. (Mckeown, 2014, 77)

Prior to Luther, celibacy tended to be venerated, but Luther prompted a shift toward esteeming marriage and procreation (Mckeown, 2014, 78).

Mckeown suggests that much of Luther's writings on marriage and procreation reflect his own personal struggle with celibacy. He points to an example from Luther's 1522 work, the *Estate of Marriage*:

> "Be fruitful and multiply," is not a command. It is more than a command, namely, a divine ordinance which it is not our prerogative to hinder or

[11] For more on the history of anti-natalism see Akerma (2021) and Lochmanová et al. (2020).

> ignore. Rather, it is just as necessary as the fact that I am a man, and more necessary than sleeping and waking, eating and drinking, and emptying the bowels and bladder. It is a nature and disposition just as innate as the organs involved in it. Therefore, just as God does not command anyone to be a man or a woman but creates them the way they have to be, so he does not command them to multiply but creates them so that they have to multiply. And wherever men try to resist this, it remains irresistible nonetheless and goes its way through fornication, adultery, and secret sins, for this is a matter of nature and not of choice. (LW 45.18 quoted in Mckeown, 2014, 78)

Luther challenged vows of celibacy because he worried those who took them would regret them and they would become a cause for sin. Mckeown explains that Luther "was especially concerned about those who had been put inside religious houses as youths by their parents (LW 44.216), and also young adults who joined impetuously (LW 39.296)" (Mckeown, 2014, 82–83). Indeed, in Monastic Vows he claims that all vows made in such contexts are invalid (LW 43.87).). Luther also worried that that idle daily life in a monastery could lead to sin. Avoiding marriage runs a great risk of leading to sexual sins (Mckeown, 2014, 83). Finally, Luther also appears to the have thought that child-rearing was a kind of penitential discipline.

Mckeown invites the reader to consider Luther's advocacy of procreation within the context of sixteenth century Europe, including that "over a third of infants died before the age of five (Ozment, 101), and, indeed, there was significant mortality among older children, young adults, and at all ages. Before and during Luther's formative early years, the population of Europe was lower than it had been two hundred years earlier" (Mckeown, 2014, 103).

Augustine

Prior to converting to Christianity, Augustine was a Manichean. Mani, the founder of that religion was an anti-natalist and the movement promoted it (see Häyry & Sukenick, 2024, 12). Mani, "deprecated reproduction because conceiving a baby imprisoned a fragment of divinity, a soul, in flesh (AE 239 citing *mor.* 2.18). There were two levels of Manichean: the 'elect' were celibate, while the married 'hearers' were encouraged to practice contraception" (Mckeown, 2014, 149–150). Augustine was perpetually fearful of accusations of being a secret Manichean, and Mckeown argues that "[i]n such circumstances it is remarkable that Augustine dared to write anything less than what was entirely positive about marriage and marital reproduction. That he sometimes did compare it unfavorably with celibacy suggests he was confident he could show that his ideas were firmly anchored in the New Testament and mainstream Christian tradition" (Mckeown, 2014, 151–152). Augustine originally accepted an interpretation in early church thinking that the first human beings created were immortal and did not reproduce, but later says that reproduction was 'original' and so occurred before the fall (Mckeown, 2014, 155).

2.5 Historical Precedents Outside of the Christian Tradition

To conclude this chapter, I briefly examine some historical precedents for anti-natalism not located centrally within the Christian tradition.

A Socio-Historical Thesis

In their short book, *Antinatalism, Extinction, and the End of Procreative Self-Corruption*, Matti Häyry and Amanda Sukenick argue for a provocative historical thesis:

> The prevailing consensus has been that antinatalist sentiments have been gradually brought to light by ever-evolving forms of Western philosophy. While this may be partially true, the wider truth is quite the opposite. The basic tenets of antinatalism have been known to thinking individuals for millennia, but every time they have threatened to disturb hegemonic establishments they have been crushed by philosophical systems such as those of Plato, Augustine of Hippo, and others. This is a bold hypothesis,

> but we believe that we can lend it some surprisingly credible support. (Häyry & Sukenick, 2024, 2)

Häyry and Sukenick assume that metaphysical naturalism is true and therefore that there is no objective meaning in life. There is no cosmic order or afterlife. They believe that procreation has become a way for humans to deceive themselves into thinking life is meaningful when it is not. They trace a long historical journey to show that anti-natalism has always existed. Instead of evaluating their arguments, I simply observe that for the reader that this is in fact a case that has been made.

Ancient Greeks

Häyry and Sukenick quote Plutarch's record of Silenus's answer to King Midas about whether life is meaningful:

> Ephemeral offspring of a travailing genius and of harsh fortune, why do you force me to speak what it were better for you men not to know? For a life spent in ignorance of one's own woes is most free from grief. But for men it is utterly impossible that they should obtain the best thing of all, or even have any share in its nature (for the best thing for all men and women is not to be born); however, the next best thing to this, and the first of those to which man can attain, but nevertheless only the second best, is, after being born, to die as quickly as possible. (Plutarch quoted in 1928, 181 in Häyry & Sukenick, 2024, 5)[12]

Sophocles writes that "[n]ot to be born is, beyond all estimation, best; but when a man as seen the light of day, this is next best by far, that with utmost speed he should go back from where he came" (quoted in Lochmanová, 2020, 40). Euripides writes, "Where a man is born we should assemble only to bewail. His lot in coming in so much evil. But when dies and comes to the end of trouble then we should rejoice and praise and his happy departure" (quoted in Lachmanová, 2020, 40). Heraclitus says, "When men are born they are fain to live and

[12] As cited by Gerber (1999). See also Svoboda (2020, 59).

suffer death, or rather go to their rest, and they leave children who also suffer death" (quoted in Lachmanová, 2020, 41). Herodotus and Empedocles also may have expressed anti-natalist sentiments (Lachmanová, 2020, 41). Finally, the Greek cynic philosophers expressed anti-natalist tendencies (e.g., Diogene) (Häyry & Sukenick, 2024, 9).

Roman Stoics

The Roman stoics also appeared to have endorsed anti-natalism, or something close to it. Consider that Cicero writes: "[N]ot being born and hence hurled against life's rocks is clearly the best option. Second best, if you do get born, is to die and flee the violence of Fortune as fast as you'd run from a burning building" (quoted in Häyry & Sukenick, 2024, 9). A hundred years later, Seneca announces similar sentiments, "Nothing is so deceptive, nothing is so treacherous as human life; by Hercules, were it not given to men before they could form an opinion, no one would take it. Not to be born, therefore, is the happiest lot of all and the nearest thing to this, I imagine, is that we should soon finish our strife here and be restored again to our former rest" (quoted in Häyry & Sukenick, 2024, 9).

Medieval Anti-Natalism

Häyry and Sukenick point to a fascinating case of medieval Arab Anti-Natalism located in the work of al-Ma'arri, a scholar who was blinded at a young age, who had this to say about procreation:

> Whenever I reflect, my reflecting upon what I suffer only rouses me to blame him that begot me. And I gave peace to my children, for they are in the bliss of nonexistence which surpasses all the pleasures of this world. Had they come to life, they would have endured a misery casting them to destruction in trackless wildernesses. (al-Ma'arri quoted in Häyry and Sukenick, 2024, 17)

Häyry and Sukenick also write that "[w]hile al-Ma'arri represented free-thinking atheism, two early second millennium South European sects, the Bogomils in the Balkans and the Cathars in Italy and Spain, carried forward the torch of Christian antinatalism" (Häyry & Sukenick, 2024, 17).

Hinduism and Buddhism

Ken Coates[13] discusses anti-natalist sentiments in Hinduism (the concept of Moksha) and Buddhism (Nirvana). It is far too extreme to say these religions are anti-natalist, but it is fascinating that anti-natalist thinking can be gleaned from them. Coates explains that "[a]mong world religions Hinduism and Buddhism stand out in their strongly negative view of existence. Liberating human beings from their bondage to earthly existence has been their chief concern" (Coates, 2014, 19). Both affirm that the goal of human existence is to transcend suffering and evil.

Of Hinduism, Coates writes that:

> According to classical Hindu beliefs, as articulated in the Upanishads and the Gita, existence in general and human existence in particular is fundamentally flawed. From the beginning to end, i.e. from birth to death, life involves suffering. Moreover everything in the phenomenal world is transitory - ephemeral and passing. Human consciousness or the soul finds itself 'trapped' in material existence and dragged through an endless cycle of birth, decay, old age and death with apparently no rhyme or reason and with no end to the process in sight. (Coates, 2014, 27)

Coates suggests that the Buddhist doctrine of Nirvana has similar implications to Moksha.

[13] This work was originally published under this pseudonym, but later the author revealed himself as Ramesh Mishra.

The Modern Period

The modern period contains numerous thinkers who express anti-natalist sentiments even they do not outright endorse it. For example, Montesquieu writes that "we should bewail people when being born and not after their death" (1989, 116 quoted on Lochmanová, 2020, 93). Anti-natalism can be found outside of the Western tradition in the work of Vladimir Sergeyevich Solovyov (1853–1900), even though he did not explicitly identify as an anti-natalists[14] (see Poledníková, 2020). In the early twentieth century, a German writing under the pseudonym of Kurnig was quite possibly the first full-fledged modern anti-natalist.[15] This writer a promoted a view called 'neo-nihilism' which is equivalent to anti-natalism. Kurnig writes that "With increasing intelligence, mankind comes to realise that, all in all, suffering far outweighs pleasure, that it must stop procreation and must do so as soon as possible. Thus: NEO-NIHILISM" (Kurnig, 1903 quoted in Akerma, 2020, 126). In addressing procreators, Kurnig writes:

> You think you're saying and doing something pretty strong, beautiful, full of character, don't you? But do you know what it is? Weakness of character and ignorance. I mourn the creatures you bring into the world who could not defend themselves when you created them, who otherwise would have protested out loud against your action. Since it all boils down to the suffering and destruction. Our race serves nothing and exists only as a result of those who, like you, do not examine things thoroughly. Life is suffering; to abstain from procreation is philanthropy and duty. (Kurnig, 1903 quoted in Akerma, 2020, 126–127)

Kurnig also seems to think that social circumstances make it difficult for individuals to admit they do not want to have children. He writes that "The silence of some of us may not confuse us. Because of external circumstances many are not allowed to admit that they are pessimists and, therefore, not prepared to have children" Kurnig quoted on Akerma, 2020, 127).

---

14 As did neither Schopenhauer nor Flaubert (Poledníková, 2020, 111).

15 Some of speculated that this writer was a medical doctor.

Alexandar von Humboldt (1769–1859) alleged memoirs contain that following remarks: "I am not cut out to be a family man. I also believe that marriage is a sin, and the production of children a crime. Whoever marries with the intention to procreate is a sinner because gives life a children without being able to give them the certainty of happiness." (Humboldt, 1894, 209n, on Akerma, 2020, 128). Edmond de Goncourt (1822–1896) and Jules de Goncourt (1830–1870), of whom established the Prix Goncourt French literary prize wrote: "How is it that in no epoch of history, in no place on earth, a sect of wise men has been formed with the aim of making human life die out in the face of cruelty of its evils? Why is it that this end of mankind by abstention from procreation has not been preached?" (Edmond and Jules de Goncourt, 1868, 279 on page Akerma 129; translation by author of chapter). The German theologian David Friedrich Strauss (1808–1874) provocatively argued that there are no longer any Christians "[b]ecause Christianity – which originally was pessimistic and sceptical about reproduction against the background of an imminent end of the world – had long since been coloured optimistically by the subliminal continuing effect of Jewish beliefs, there were actually no more real Christians..." (Strauss quoted in Akerma 133).

Existentialism and Absurdist Literature

Sarte, Camus, Unamuno, Cioran all anticipate anti-natalist sentiments (though only Cioran is explicitly an anti-natalist). Most discuss existence and somehow reconcile themselves to it (Coates, 2014, 71). Peter Wessel Zapffe (1899–1990) was a Norwegian existentialist whose main work *On the Tragic* (1941) is still only available in Norwegian. However, "Zapffe who must be credited with being the first rejectionist to come up with the idea of anti-natalism as the way out of existence for humans" (Coates, 2014, 72). Coates claims that Zapffe anticipates Benatar when he writes that:

> Quite apart from the need for an overarching meaning of life humans have developed values and sentiments absent in all other species. Longing

> for a just world is one of them, compassion for the suffering of all sentient beings is another. Yet these have no place in the universe. The lack of meaning reveals to man, 'a nightmare of endless repetition, a senseless squander of organic material' while the 'suffering of human billions makes its entrance into him through the gateway of compassion'. (Zapffe quoted in Coates, 2014, 73)[16]

According to Coates, philosophers such as Schopenhauer, Nietzsche, and Kierkegaard were some of the first to engage directly with existential issues, including the broad themes concerning the nature of human existence (Coates, 2014, 39). He claims that:

> On the evils of existence: Schopenhauer emphasizes pain and suffering as the defining characteristics of life. And although the pointlessness and futility of existence is a part of his conception of this 'evil' it receives far less attention and emphasis. It is the moral rejection of the world that he emphasizes rather than its metaphysical worthlessness as a contingent phenomenon. (Coates, 2014, 57)

There are modern literary expressions of anti-natalism to be found in the absurdist literary tradition, including in the work of Samuel Beckett and Jean Paul Sartre, especially the former's novel *Nausea.* Literary expressions have not been the focus of this chapter, nor will they be in what follows. Yet again, however, it shows that anti-natalist thinking has been around for longer might be supposed. Coates describes this branch of literature as endorsing *rejectionist* philosophy:

> Rejectionism does not believe in the calculus of pain and pleasure not only because any such exercise is impossible given that there is no common unit of measurement but also because its moral condemnation of existence is based on the irremediable presence of 'evil' in the world. It follows that to endorse existence is to condone evil, indeed to invite evil, albeit unintentionally. It follows that those who support and endorse existence are responsible, even if indirectly, for the crimes of humanity. To summarize: the rejectionist viewpoint has a long history stretching over

16 See also Akerma (2020, 128).

> millennia and a core of basic beliefs. Justifiably then rejectionism may be identified as a distinct attitude to life, a more or less coherent worldview. (2014, 126–127)[17]

2.6 Conclusion

The purpose of this chapter was to show that anti-natalism does not emerge for the first time with the work of analytic philosophers such as Benatar, among others. I wanted to show that there is at least some precedent in the Christian tradition for anti-natalist thinking, if not for a full endorsement of anti-natalism. I accomplished this by explaining alternative interpretations of pro-natalist Bible passages (2.2), before turning to explore sections of the Bible that could be construed as anti-natalist (2.3). I then considered some influential figures in the history of Christianity, and the various motivations they may have had for either pro-natalist or anti-natalist views (2.4). I concluded by highlighting some anti-natalist sentiments outside of the Christian tradition, including ideas located in Buddhism, Hinduism, the Modern period, and Existentialist writers (2.5). Anti-Natalist thinking did no emerge in the vacuum in contemporary analytic philosophy.

This chapter is admittedly haphazard in its coverage. Indeed, the historical scholarship on anti-natalism remains rather unsystematic and uncomprehensive. My aim was not to give a comprehensive history of anti-natalism within Christian thought, nor was it to defend a particular interpretation of certain Biblical passages or Christian history. Instead, I wanted to show that there are plausible *alternative* interpretations of the fruitful verses, in addition to the fact that it is possible anti-natalist thinking has been part of the Christian tradition in various forms for quite some time. Pro-natalism cannot be assumed as the *appropriate* default stance of the Chrisitan tradition, though I have little doubt that it is in fact the default stance.

---

[17] See https://www.exploringantinatalism.com/antinatalismquotes/ which contains an extensive list of anti-natalist quotes (though I do not think every quote listed actually points to genuinely anti-natalist sentiments of the author).

To conclude, this chapter should be viewed as an invitation to Biblical scholars, theologians, and historians of Christianity to consider more carefully whether there are anti-natalist sentiments throughout parts of the Christian tradition. If the ideas I have presented in this chapter are entirely misguided, it is up to others with more familiarity with the relevant disciplines to correct them. In showing that anti-natalism has perhaps been a viable, albeit minority, position for Christians, I hope to have softened the reader to the arguments for Christian anti-natalism that I will make later. Finally, it is important to remember that just because there are anti-natalist sentiments within the Christian tradition, this does not in itself show that Christian anti-natalism is true. Alternatively, if there were no Christian anti-natalists or if all of the precedents in this chapter are mistaken, it would also not on its own show that Christian anti-natalism is false. What is needed to show that anti-natalism is true and that it could also be compatible with Christianity is philosophical argument. This is the task I begin to take up in the next chapter, where I provide a critical overview of the main arguments for anti-natalism that are offered by David Benatar, while also noting which ones are *prima facie* consistent with Christianity.

Part II

The Contemporary Philosophical Debate

3

Benatar's Arguments for Anti-Natalism

3.1 Introduction

The purpose of this chapter is to offer an overview of the arguments for anti-natalism located in the work of David Benatar, a leading contemporary proponent of anti-natalism. The arguments from Benatar that I will explain are the Asymmetry Argument (3.2), the Quality of Life Argument (3.3), and the Misanthropic Argument (3.4).[1] After outlining each argument, I briefly highlight some of the main objections and replies in the literature. However, given the now vast literature on Benatar's arguments, I make no attempt at completeness, nor do I issue a verdict on the soundness of his arguments. Instead, I focus on exploring whether the arguments are at least *prima facie* consistent with Christianity, and hence could be appealed to in building a case for the claim that Christians

[1] Benatar also argues that the fact there is no cosmic meaning entails anti-natalism (2017). However, since this argument is not *prima facie* consistent with Christianity, I will not spend time explaining it here. See Metz (2022) for an objection to this argument and Lougheed (2022) for a rejoinder.

K. Lougheed, *Christian Anti-Natalism*, https://doi.org/10.1007/978-3-032-03902-6_3

should be anti-natalists. In the final section, I consider a frequent objection that has been brought against Benatar's anti-natalism which is that it entails pro-mortalism and show why it fails (3.5). This is important because both the objection and response can generalise to anti-natalism beyond Benatar's arguments for it. My main conclusion is that all three of the arguments from Benatar are *prima facie* consistent with Christianity. Though they will not be the focus of the main arguments for Christian anti-natalism in Part III, Christians wishing to avoid anti-natalism need to demonstrate where Benatar's arguments fail.

Before proceeding, it is worth noting that despite its length, this chapter is a relatively tertiary overview of Benatar's work. Indeed, the initial draft was about twice of its current length, and I had to make difficult choices about which material to include. I chose to focus on expositing the arguments in detail and then summarising what I consider to be the most substantive philosophical objections to them. I have also been guided to some extent by Benatar's replies to his critics that have appeared in various journal articles over the past decade and a half (e.g., 2012, 2013, 2022).

Finally, before moving to the details of the chapter, a comment on the scope of Benatar's arguments is in order. Benatar believes that his arguments apply to any sentient beings, though his focus is on humans (2006, 2). It has also been suggested that for an anti-natalist "[t]o be against the reproduction of extraterrestrials would, of course, be a markedly exotic stance" (Häyry's & Sukenick, 2024, 29). However, it is difficult to see how this would be so exotic for the anti-natalist whose arguments are intended to apply to *sentient* creatures. If there are extraterrestrials who are sentient and experience harm, then many of the arguments for anti-natalism will also apply to them, including some of Benatar's arguments.[2] Still, this is ultimately neither here nor there as I am concerned with whether Christians should be anti-natalists. As such I am only discussing humans and it is not relevant whether the arguments turn out to apply to none, some, or all other sentient creatures.

[2] It is also an open question whether machines made by people are sentient. If they are, then arguments for anti-natalism would also apply to them (Häyry & Sukenick, 2024, 31; see also Akerma, 2023; Metzinger, 2017).

3.2 The Asymmetry Argument

Benatar's best-known argument for anti-natalism is the Asymmetry Argument which focuses on the harms that every person brought into existence will experience. Benatar's main point is that:

> Although the good things in one's life make it go better than it otherwise would have gone, one could not have been deprived by their absence if one had not existed. Those who never exist cannot be deprived. However, by coming into existence one does suffer quite serious harms that could not have befallen one had one not come into existence. (2006, 1)

Thus, Benatar's argument does not claim that those who never exist are better off than those who exist. Non-existent beings have no referent. Rather the argument is that coming into existence is always a serious harm. Not existing is not good for anyone, but existing is bad for those who exist (Benatar, 2006, 4). Benatar assumes that a person's genetic makeup is essential to their personal identity. A person cannot have had different genetic parents and still be the same person. The chances of a particular sperm and egg forming the zygote that eventually becomes you is all extremely unlikely. It is very unlikely that *you* should have been born. In light of this, "[t]he recognition of how unlikely it was that one would have come into existence, combined with recognition that coming into existence is always a serious harm, yields the conclusion that one's having come into existence is *really* bad luck" (Benatar, 2006, 7). Even if the world were better, Benatar believes that the having children would still be impermissible. Coming into existence is always a net loss.

The Argument

The Asymmetry Argument can be motivated by reflecting on the non-identity problem. Benatar explains that "[t]he problem arises in those cases where the only alternative to bringing a person into existence with a poor quality of life is not to bring that person into existence at all"

(2006, 19).[3] The same person cannot be brought into existence without the corresponding harm. Consider the parents of a child with a genetic disorder who cannot create a child who will not suffer from the disorder. If such an impairment is essential to personal identity, then it is not wrong to bring them into existence. They are not made worse off by the impairment since they could not exist without it.

Benatar responds to this by observing that there is a difference between a life worth continuing and a life worth starting. Whether a life is worth continuing is a judgment about a being that already exists. However, whether a life is worth starting is a judgment about a potential but a non-existent being (Benatar, 2006, 22). People have conflated the former with the latter in order to claim that lives are worth starting.[4] In other words:

> When they distinguish between impairments that make a life not worth living and impairments that, though severe, are so bad as to make life not worth living, they are making the judgments in the present-life cases. Those lives nor worth living are those that would not be worth continuing. But the problem is that these notions are then applied to future-life cases. In this way, we are led to make judgements about future-life cases by the standards of present-life cases. (Benatar, 2006, 23)

It is therefore intelligible to claim that it is wrong to start a life that once started would be worth continuing (Benatar, 2006, 23).

Benatar argues it is *always* impermissible to start a life even though many people assume that bringing a person into existence is permissible if their life would be on balance good (Benatar, 2006, 28). For Benatar, *always* does not mean *necessarily*. This is because his argument does not apply to possible worlds that only contains good and never any bad. The right response to those lives is indifference (i.e., it is neither good nor bad that they came into being). However, since all lives in our world will contain at least some bad (and according to Benatar, much more

---

[3] For the non-identity problem (see Parfit, 1984, 359). For the paradox of future individuals (see Kavka, 1982, 93–112).

[4] Benatar cites Parfit (1984, 358–359) and Feinbeg (1992, 26) as two examples of this mistake.

badness than is typically acknowledged), it is always a harm to come into existence (2006, 29).

The Asymmetry argument hinges on the asymmetry between pleasure and pain. According to Benatar, "there is a crucial difference between harms (such as pains) and benefits (such as pleasures) which entails that existence has no advantage over, but does have disadvantages relative to, non-existence" (2006, 30). The key difference is the absence of pain is good even if not one exists to enjoy its absence while the absence of pleasure is not bad unless a person exists who would be deprived of it (Benatar, 2006, 30).

Consider the following four claims:

> (1) the presence of pain is bad,
> and that
> (2) the presence of pleasure is good. (Benatar, 2006, 30)

However, Benatar claims that this symmetry does not appear to apply to the absence of pain and pleasure:

> (3) the absence of pain is good, even if that good is not enjoyed by anyone,
> whereas
> (4) the absence of pleasure is not bad unless there is somebody for whom this absence is a deprivation. (Benatar, 2006, 30)

This means that "[t]he judgement made in (3) is made *with reference to the (potential) interests of a person* who does or does not exist" (Benatar, 2006, 30). This means that regarding the pain of someone who exists, it "would have been good even if this could only have been achieved by the absence of the person who now suffers it. In other words, judged in terms of the interests of a person who now exists, the absence of the pain would have been good even though this person would then not have existed" (Benatar, 2006, 31).

What Benatar calls the Basic Asymmetry (i.e., the asymmetry between (3) and (4)) has broad explanatory power in being able to explain at least four other procreative asymmetries:

- *The asymmetry of procreational duties* (Benatar, 2013, 123): There is a duty to avoid bringing someone who will suffer into existence but there is no corresponding duty to bring happy people into existence (Benatar, 2006, 32).
- *The prospective beneficence asymmetry* (Benatar, 2013, 123): It is inappropriate to say that the reason for having a potential child is that they will be benefitted by being brought into existence. However, "it is not strange to cite a potential child's interests as a basis for avoiding bringing a child into existence" (Benatar, 2006, 34).
- *The retrospective beneficence asymmetry* (Benatar, 2013, 123): While both bringing people into existence and failing to do so can be a source of regret, "only bringing people into existence can be regretted *for* the sake of the person whose existence was contingent on our decision" (Benatar, 2006, 34). Those who are not brought into existence never existed and so cannot be regretted for their sakes. It is intelligible to regret not having had children, but not for the children's sake – for they never existed. Instead, the disappointment is over missing out on the relevant parental experiences (Benatar, 2006, 35).
- *The asymmetry of distant suffering and absent happy people* (Benatar, 2013, 123): Consider the suffering of distant people compared to uninhabited portions of earth. Benatar claims that "we rightly are sad for inhabitants of a foreign land whose lives are characterized by suffering, when we hear that some island is unpopulated, we are not similarly sad for the happy people who, had they existed, would have populated this island" (Benatar, 2006, 35). Consider that we do not mourn for the people who do not exist on Mars (Benatar, 2006, 35).

In light of the asymmetry between pleasure and pain, Benatar argues that coming into existence is always a harm. This can be further illustrated by the following Figure[5].

[5] Modified from Benatar (2006, 38) and Benatar (2013, 124).

If X exists (Scenario A)	If X never exists (Scenario B)
(1) Presence of pain (Bad)	(3) Absence of pain (Good)
(2) Presence of pleasure (Good)	(4) Absence of pleasure (Not bad)

The absence of pain is better than the presence of pain. The presence of pleasure is not better than the absence of pleasure (Benatar, 2006, 38). The absence of pleasure is not a privation for X since X does not exist. But if X exists, then it would be better to have pleasure than to be deprived of it. On the other hand, the absence of pleasures, "by contrast, are not neutral states of some person. They are no states of a person at all" (Benatar, 2006, 41). There is no one that is worse off for not having experienced the pleasure (Benatar, 2006, 42). Benatar's Asymmetry implies that "[t]he fact that one enjoys one's life does not make one's existence better than non-existence, because if one had not come into existence there would have been nobody to have missed the joy of leading that life and thus the absence of joy would not be bad" (2006, 58).[6]

To further explain why (2) is not better than (4), Benatar offers the following analogy:

> S (Sick) is prone to regular bouts of illness. Fortunately for him, he is also so constituted that he recovers quickly. H (Healthy) lacks the capacity for quick recovery, but he *never* gets sick. It is bad for S that he gets sick and it is good for him that he recovers quickly. It is good that H never gets sick, but it is not bad that he lacks the capacity to heal speedily. The capacity for quick recovery, although a good for S, is not a real advantage over H. This is because the absence of that capacity is not bad for H. This, in turn, is because the absence of that capacity is not a deprivation for H. H is not worse off than he would have been had he had the recuperative powers of S. S is not better off than H in any way, even though S is better off than he himself would have been had he lacked the capacity for rapid recovery. (2006, 42)

[6] Benatar also claims that the basic asymmetry "solves the non-identity problem, avoids the repugnant conclusion and avoids the mere addition problem (Benatar, 2013, 140; see also Benatar, 2006, 168–178).

This is not intended to show that (2) is good and (4) is not bad. Instead, it shows that (2) is not better than (4), and (1) is worse than (3). So, Scenario B is to be preferred to Scenario A (Benatar, 2006, 43).

Elsewhere Benatar writes that:

> The basic asymmetry strikes me as a fundamental moral truth. I suspect that it is widely accepted – that is, until people see where it leads, namely to the conclusion that coming into existence is always a harm. Once people see this implication they scramble desperately to find some way to avoid having to accept the asymmetry. (Benatar, 2012, 129)

It is to some of these objections that I turn to in the next section.

Objections and Replies

It is probably safe to say that Benatar's Asymmetry Argument is by far the most widely discussed argument in the contemporary philosophical literature on anti-natalism. In what follows, I briefly summarize some of the main objections and replies to the argument.

Elizabeth Harman is one of the few who has outright denied the asymmetry between (3) and (4). Specifically, she writes that "[w]hen we fail to bring someone into existence who would have suffered horribly, then we have failed to do *something we should not have done.* We have acted rightly, and we have avoided causing a bad state of affairs. None of that implies that there is anything *good* about the current situation" (Harman, 2009, 781) In other words, the absence of pain is not good, even though we acted rightly. However, Benatar never denies that it is logically consistent to make such a claim. His point is axiological: we should not deny it (Benatar, 2013, 126).

Chris Kaposy denies (3) by arguing that there is no positive value in the absence of suffering. Instead, the positive value comes from not being the cause of the suffering (Kaposy, 2009, 105; see also Benatar, 2013, 128). People have a duty to avoid being the cause of pain. The absence of pain is not good, it is perhaps indifferent (Kaposy, 2009, 106; see also Benatar, 2013, 128). Benatar counters that "(3) does not claim that the absence of pain is intrinsically good. It [i.e., the Asymmetry] claims that it is better than the presence of pain—that is, it is comparatively good. Denying (3) is thus not only an unusual view, but also an extreme one" (2013, 128).

David DeGrazia argues that asymmetries (iii) and (iv) are not explained by the basic asymmetry between (3) and (4).[7] This is because non-existent beings do not exist and so cannot have interests. This presumably includes the interests of not being harmed. He asserts the contrasting asymmetry that, "[r]egarding those who exist, the presence of benefit is good and the presence of harm is bad; regarding those who never exist, the absence of benefit and the absence of harm are neither good nor bad. Only those who exist have interests and can be benefited or harmed" (DeGrazia, 2010, 324). Benatar responds that even if DeGrazia is correct, his view still entails anti-natalism (2013, 131).

Ben Bradley argues that the basic asymmetry is incoherent (Bradley, 2010, 2). This is because Benatar believes that in Scenario A the presence of pleasure is good but not better than the absence of it in scenario B. Thus, in Figure, though quadrant (2) is good is not better than quadrant (4). Benatar responds by supposing, for the sake of argument, that (2) is not better than (4) is indeed incompatible with the two above accounts of good (2013, 135). But he claims that this does not show that the basic asymmetry is mistaken. Benatar responds that "it would hardly be surprising if extant accounts of good had not considered the unusual case of comparisons between existing and never existing" (2013, 135). This is exactly why the non-identity problem arises; people attempt to view procreational harm by employing an understanding of harm that originally arose outside of discussions of procreation (Benatar, 2013, 135).

Thaddeus Metz argues that the Basic Asymmetry does not explain (i) the asymmetry of procreational duties. According to Metz, "the judgment that we have a duty not to create suffering people is at least equally well explained by the principle that *it is permissible to start a life if and only if it would be worth continuing*" (2011, 241). That it is permissible does not entail that there is a requirement to procreate. Benatar responds

[7] DeGrazia also appeals to the work of Jeff McMahan in claiming that (i) the asymmetry of procreational duties might be basic or fundamental (2010, 322; see McMahan, 1998). Bayne (2010, 51) raises the same objection. See Benatar (2013, 130) for a response.

that this is not an alternative explanation of the Basic Asymmetry; it is merely a restatement of it (2012, 133).

Rivka Weinberg argues that the four procreative asymmetries are better explained by the common moral fact that "our moral obligations are to persons who do or will, i.e., to persons with moral standing" (2012, 28).[8] Non-existent, hypothetical entities, have no moral standing. There are no duties to such non-existent persons. Alternatively, they can also be explained by the basic metaphysical fact that "[a]ll interests are contingent upon existence" (Weinberg, 2012, 28).[9] Benatar responds that Rivka is selective in choosing when to invoke the metaphysical and moral fact (Benatar, 2012, 136).

Some other objections include Skott Brill's challenge to the analogy of Sick and Healthy (2022). However, even if such challenges succeed, Benatar has always been clear that he is not relying on the analogy to prove his conclusion (2012, 141–143). David Boonin attempts to reject the Basic Asymmetry by providing two alternative principles purporting to better explain asymmetries (i) to (iv) (2012).[10] Benatar acknowledges that they explain (i) the asymmetry of procreational duties contra Boonin, but he denies that it explains (ii) to (iv) (Benatar, 2012, 138). Brian McLean argues that (i) to (iv) ought to be explained in terms of "the generation of reasons rather than an asymmetry in terms of what's good or bad for someone" (2015, 123). This allows McLean to appeal to the idea of a 'disabler', "the concept of a consideration that prevents another consideration from being a pro tanto moral reason for or against some action" (2015, 123–124). Oliver Hallich claims that non-existent persons can be harmed by not being brought into existence, though Bentar says Hallich misunderstands the argument (2022, 120–123). Deke Caiñas Gould argues that the possibility of non-human-like minds who aren't harmed by coming into existence poses a challenge to the Asymmetry Argument for ant-natalism (2021). I have responded by suggesting the mere possibility of such minds does nothing to undermine

[8] Hare argues that moral concern ought to be shown for those who do not exist (2007). See Wienberg (2012) for a response.

[9] Weinberg believes that the metaphysical fact grounds the moral fact, but this claim is not necessary for her argument (2012, 28).

[10] See Smuts (2014) for a related discussion.

the basic asymmetry (Lougheed 2023). Finally, Fumitake Yoshizawa argues that either the Basic Asymmetry cannot explain asymmetries (i) to (iv) or it does not entail anti-natalism (2021).

Prima Facie Consistency with Christianity

Supposing the Asymmetry Argument survives these challenges, it is important to consider whether there is any initial reason why a Christian could not accept it (beyond doubting the supposition that its conclusion is incompatible with Christianity). I will address this question in more detail in Chapter Five, but there could be arguments that can be eliminated from consideration without a deeper discussion. On the face of things, I see no reason why the Asymmetry Argument should be disqualified from being considered by the Christian. Christians ought to be seriously concerned by preventing harm to others. The focus of the Asymmetry Argument is on eliminating harm and this is something Christians should be pursuing, even if ceasing to procreate is hardly thought of as a plausible way of achieving this end. Of course, there are larger questions and objections about whether a Christian could endorse this argument (e.g., does it require accepting positions a Christian will probably want to reject such as negative utilitarianism or welfarism). My point is just that there is nothing obviously inconsistent with Christianity in the Asymmetry Argument. Remember that this just consists of the Basic Asymmetry, along with the claim that it serves as the best explanation of the other four asymmetries.

3.3 The Quality of Life Argument

The Asymmetry Argument implies that if a potential life were to contain any harm at all, even if very slight, it is better that life not become actual. But even if one denies that slight harms would make a life not worth starting, Benatar suggests that most will recognise a life containing a lot of harm is not worth starting. The Quality of Life Argument trades on the fact that human life contains much more harm and suffering than most people realise. This means that "[i]f people realized just how bad their lives were, they might grant that their coming into existence would have been a harm had their lives contained but the smallest amount of bad" (Benatar, 2006, 60). Still, the Quality of Life Argument constitutes a separate argument for anti-natalism.[11] Indeed, Benatar suggests that though most lives may be worth continuing, they are typically much worse than people realise, containing a significant amount of harm (Benatar, 2006, 61).

The Argument

Benatar begins his description of the argument by claiming that the quality of a life cannot be assessed merely by subtracting the negative value it contains from the positive value. That methodology would mean assigning values to quadrants (1) and (2) in Figure and subtracting them from (3) from (4). However, the distribution, among other considerations, also matters in quality of life assessments (Benatar, 2006, 61). The *order* matters such that a life where things are quite good then turn quite bad is worse than one where the good and bad is more spread out. This is so even if the overall goods and bads in the two lives are of a similar quality (Benatar, 2006, 62). The *intensity* of the goods and bads matter, in addition to the *length* of life (Benatar, 2006, 62–63).

There are also relevant non-distributional assessments that matter. Benatar suggests that if a threshold for badness is reached there is no amount of goodness that could subsequently outweigh that badness (Benatar, 2006, 63). Here Benatar appeals to the story of Donald Cowart, which has become a well-known case study in bioethics:

[11] The argument can also be understood as a bolstering the Asymmetry Argument, which does not discuss the quality or quantity of harms that people experience. Though every person is harmed by being brought into existence, the degree to which they are harmed depends on the quality of their particular life. Benatar argues that all lives contain significant amounts of bad (Benatar, 2006, 61). See Erik Magnusson for a suggestion that the Asymmetry Argument relies on the Quality of Life Argument in important ways not recognised by Benatar (2019). Examining this claim would take me too far afield as the relationship between these arguments is not important for my purposes.

> It is just this assessment that Donald ('Dax') Cowart made of his own life—or at least of that part of his life following a gas explosion that burnt two-thirds of his body. He refused extremely painful, life- saving treatment, but the doctors ignored his wishes and treated him nonetheless. His life was saved, he achieved considerable success, and he reattained a satisfactory quality of life. Yet, he continued to maintain that these post-burn goods were not worth the costs of enduring the treatments to which he was subjected. No matter how much good followed his recovery, this could not outweigh, at least in his own assessment, the bad of the burns and treatment that he experienced. (2006, 63)[12]

Benatar also suggests that a life containing a lot of lows and a lot of highs might be worse than a life with fewer highs and lows, even if the former life has more positive value overall (Benatar, 2006, 63–64).

Psychological Data

Benatar explains that most people do not believe that their lives are all things considered bad (2006, 64). In fact, "most people think that their lives go quite well" (Benatar, 2006, 64). These self-assessments are frequently considered to be strong evidence against the claim that life is bad; most people do not believe that coming into existence is a harm. However, Benatar appeals to psychological literature to show that such self-assessments are unreliable (2006, 64).

For example, there is a psychological tendency for individuals to overrate their abilities, in addition to overrate how well their life is going. Indeed, those with more accurate views of their own abilities tend to be more depressed and function less well than their peers who have more unrealistic and inflated views of their abilities (Benatar, 2006, 64–65).[13]

Pollyannaism is the psychological tendency that people have a tendency towards the positive (Benatar, 2006, 64–65).[14] This includes

---

[12] There is an interesting connection here to the problem of evil. If Benatar is correct, it seems that this would nullify theodicies responding to the problem of evil by appealing to greater goods. Or, at least it would for those evils that cross a certain threshold (see Adams, 2000).

[13] Benatar cites Taylor (1989) and Taylor and Brown (1998).

[14] Benatar cites Matlin and Stang (1978, 141–144) and Greenwald (1980).

the fact that people are more likely to remember positive experiences over negative ones (Benatar, 2006, 65). This makes assessments of how well our lives are going distorted, in addition to biasing our expectations about how well things will go in the future (Benatar, 2006, 65).[15] Benatar explains that "[t]he Pollyannaism typical of recall and projection is also characteristic of subjective judgements about current and overall well-being. Many studies have consistently shown that self-assessments of well-being are markedly skewed toward the positive end of the spectrum" (2006, 65).[16] The vast majority of people describe themselves as 'pretty happy' or 'very happy', claiming they are doing better than the average person (Benatar, 2005).[17]

Consider that items that should intuitively raise a person's quality of life do not seem to affect self-assessments. For instance, "although there is a correlation between people's own rankings of their health and their subjective assessments of well-being, objective assessments of people's health, judging by physical symptoms, are not as good a predictor of peoples' subjective evaluations of their well-being" (Benatar, 2006, 66).[18] In most countries, the self-assessments of the poor indicate they are almost as happy as the rich. Education and occupation appear to make only marginal differences in self-assessments about how well a life is going.[19] Indeed, if education does make a difference, it tends to decrease happiness.[20] Extremely negative events simply do not affect people's self-assessments in the way that seems obvious they would affect them.[21]

What Benatar refers to as 'adaptation, accommodation, or habituation' can explain at least some of Pollyannaism. This is because "When

---

[15] Benatar cites Taylor and Brown (1998, 196–197), Matlin and Stang (1978, 160–166), and Weinstein (1980, 806–820; 1984).

[16] Benatar cites Inglehart (1990, 218 ff), Andrews and Withey (1976, 207 ff, 376), Diener (1996, 181–185), and Myers and Diener (1996, 70–72).

[17] Benatar cites Campbell et al. (1976, 24–25), Matlin and Stang (1978, 146–147), and Andrews and Withey (1976, 334).

[18] Benatar cites Diener et al. (1999, 287); see also Breetvelt and van Dam (1991, 981–987). Benatar also notes that some studies show disabled persons have similar assessments to nondisabled persons. He cites Cameron et al. (1973, 207–214) and Yerxa and Baum (1986, 271–283).

[19] Andrews and Withey (1976, 138–139) and Inglehart (1990, 227–232).

[20] Benatar cites Campbell et al. (1976, 487).

[21] See Benatar cites Diener et al. (1999, 286–294).

a person's objective well-being takes a turn for the worse, there is, at first, a significant subjective dissatisfaction. However, there is a tendency then to adapt to the new situation and to adjust one's expectations accordingly" (Benatar, 2006, 67).[22] Admittedly, there is disagreement about how much adaptation occurs and the extent to which it may vary from domain to domain. But the fact remains that subjective well-being tracks recent changes in well-being better than actual well-being which indicates that it is unreliable (Benatar, 2006, 68).

In sum, "of these three psychological phenomena, it is only Pollyannaism that inclines people unequivocally towards more positive assessments of how well their life is going" (Benatar, 2006, 68). Humans adapt to both negative and positive situations, comparing ourself to those in worse but also in better situations. However, "given the force of Pollyannaism, both adaptation and comparison operate both from an optimistic baseline and under the influence of optimistic biases" (Benatar, 2006, 68).

Three Theories of the Quality of Life

After outlining the psychological literature in order to show the self-assessments about one's quality of life are unreliable, Benatar turns to explaining that life goes poorly on any of the three main philosophical theories of quality of life.

According to hedonistic theories of quality of life, the main thing that matters is mental states. The more positive mental states that one experiences (factoring in both quality and quantity), the better a life goes. Benatar explains that "[n]egative mental states include discomfort, pain, suffering, distress, guilt, shame, irritation, boredom, anxiety, frustration, stress, fear, grief, sadness, and loneliness" (Benatar, 2006, 70). Positive mental states include relief from negative mental states and also intrinsically positive states which "include pleasant sensory experiences—tastes, smells, visual images, sounds, and tactile sensations—as well as some

22 Benatar cites Campbell el al. (1976, 163–164, 485), Brickman et al. (1978, 917–927), Headey and Wearing (1989, 731–739), and Suh et al. (1996, 1091–102). See also Diener et al. (1999, 285–286).

non-sensory conscious states (such as joy, love, and excitement). Some pleasures have both relief and intrinsic components" (Benatar, 2006, 71).

Benatar suggests that Pollyannaism explains why we tend to focus much more on positive mental states over negative ones. Consider that these negative states:

> [I]nclude hunger, thirst, bowel and bladder distension (as these organs become filled), tiredness, stress, thermal discomfort (that is, feeling either too hot or too cold), and itch. For billions of people, at least some of these discomforts are chronic. These people cannot relieve their hunger, escape the cold, or avoid the stress. However, even those who can find some relief do not do so immediately or perfectly, and thus experience them to some extent every day. (Benatar, 2006, 71)

We typically experience some of these states throughout much of the day. One is often tired, or thirsty, or hungry, too hot or too cold (Benatar, 2006, 72). Benatar acknowledges that we do not typically reflect on how much often we have these negative states and not only does Pollyannaism explains this but "Adaptation also plays a role. People are *so* used to the discomforts of daily life that they overlook them entirely, even though they are so pervasive" (Benatar, 2006, 72). This makes such states unlikely to factor into subjective assessments of well-being.

Benatar adds that the negative states he just mentioned are the baseline of a healthy, normally functioning human life. But chronic illness and the aches that come with ageing are a part of life too. Other states experienced by some people abound, including "allergies, headaches, frustration, irritation, colds, menstrual pains, hot flushes, nausea, hypoglycaemia, seizures, guilt, shame, boredom, sadness, depression, loneliness, body-image dissatisfaction, the ravages of AIDS, of cancer, and of other such life-threatening diseases, and grief and bereavement" (Benatar, 2006, 72).

According to Benatar, these ideas are also relevant to desire-fulfillment theories of the quality of life inasmuch as one desires positive mental states and seeks to avoid negative ones (Benatar, 2006, 73). The prevalence of negative mental states seems to indicate that many of us have unfulfilled desires. This implies that "there is a lot of dissatisfaction and not that much satisfaction, contrary to what many people think" (Benatar, 2006, 73). Still, desire-fulfillments are different because a desire could be fulfilled within a corresponding positive mental state. Likewise, one can be mistaken about whether a desire has in fact been fulfilled (Benatar, 2006, 74). In light of this, there is "further scope for error in self-assessments of well-being. Given Pollyannaism, it is evident that this error will tend towards *over*estimation of how good life is on the desire-fulfilment view" (Benatar, 2006, 74).

In fact, Benatar argues that our lives tend to be characterised by unsatisfied desires:

> One waits still longer to get rest when one is tired. Children wait years to gain independence. Adolescents and adults can wait years to fulfil desires for personal satisfaction or professional success. Where one's desires are fulfilled, this fulfilment is often ephemeral. One desires public office and is elected but not re-elected. One's desire to be married is eventually fulfilled, but then one gets divorced. One wants a holiday but it ends (too soon). Often one's desires are never fulfilled. One yearns to be free, but dies incarcerated or oppressed. One seeks wisdom but never attains it. One hankers after being beautiful but is congenitally and irreversibly ugly. One aspires to great wealth and influence, but remains poor and impotent all one's life. One has a desire not to believe falsehoods, but unknowingly clings to such beliefs all one's life. Very few people ever attain the kind of control over their lives and circumstances that they would like. (2006, 75)

One also desires not to lose what one has such as youth, family, and close friends. If a desire is fulfilled, a new one often immediately arises (Benatar, 2006, 75–76). Additionally, many desires, especially pleasures are achievements states, not default states.

Benatar also observes that the above discussions also apply to objective list theories of the quality of life. This is because any list of objective goods must include the existence of pleasure mental states, along with the absence of negative or pain states. It also must include the fulfillment of at least some desires. Benatar suggests that in the very same way humans adapt to their circumstances, objective lists are developed to include items such that at least some people can be said to be flourishing (Benatar, 2006, 81).

Notice that objective list theories are not truly objective; they are created from a human perspective. They are only objective in the sense that they purport not to vary from person to person (Benatar, 2006, 80–81). This means that if you want to know just how good human lives are, the human perspective is entirely unreliable given the psychological data above. In other words, "what we take to be worthwhile is very much determined by the limits of what we can expect" (Benatar, 2006, 82). For example, because no one lives until the age of 240, we do not think it is objectively bad not to reach that age. But Benatar writes that "most people regard it as tragic when somebody dies at forty (at least if that person's quality of life was comparatively good). But why should a death at ninety not be tragic if a death at forty is? The only answer can be that our judgement is constrained by our circumstances" (Benatar, 2006, 82). Once this is recognised, Benatar claims it shows that most human lives are bad.

Benatar claims that all three views on the quality of life can distinguish between:

a) how good a person's life actually is, and
b) how good it is thought to be.

This distinction even applies to hedonism because "the hedonistic view says that life is better or worse to the extent that it is *actually* characterized by positive or negative mental states. Since people can be mistaken about that, the hedonistic view does indeed allow the distinction between (a) and (b)" (Benatar, 2006, 87). Of course, (a) and (b) can interact such that if one mistakenly thinks their very bad life is actually good, this is at least one way in which their life is better than it would be otherwise (Benatar, 2006, 87).

In light of the evidence that people misjudge how well their lives are going such that most people's lives are going far worse than they imagine, Benatar suggests that "we are in a better position to determine whether starting such a life is indecent, given that starting a new life can never benefit the person whose life is begun" (2006, 87). With this in mind,

Benatar suggests that follow test to determine whether in the light of the typical quality of life that procreation is justified: "ask whether the amount of harm in a life could decently be inflicted on an already existent being, but neither to advance that being's overall interests nor for utilitarian purposes" (Benatar, 2006, 88). A person's own interests must be excluded because coming into existence can never benefit the person who comes into existence. Utilitarianism does not favour procreation either since new people may bring benefits to other persons, "but these are modest benefits to other individuals [...] rather than the maximizing benefits of utilitarianism" (Benatar, 2006, 88). Benatar has not argued that most lives are so bad that they are not worth continuing but instead that most lives contain a high degree of badness (Benatar, 2006, 88).

Benatar recognises that some may not accept the conclusions above about the quality of most lives. Perhaps the ordinary mental states involving hunger, thirst, aches, and pains do not way heavily on them. But even setting this aside, Benatar claims that the world contains a great amount of 'unequivocal' suffering (2006, 89). Consider the following:

> *Natural Disasters*
> More than fifteen million people are thought to have died from such disasters in the last 1,000 years. In the last few years, flooding, for example, has killed an estimated 20,000 annually and brought suffering to 'tens of millions'. The number is greater in some years. In late December 2004, a few hundred thousand people lost their lives in a tsunami. (Benatar, 2006, 89)

> *Hunger and Malnutrition*
> Approximately 20,000 people die every *day* from hunger. An estimated 840 million people suffer from hunger and malnutrition without dying from it. (Benatar, 2006, 89–90)

> *Disease*
> Disease ravages and kills millions annually[...]. The 1918 influenza epidemic may have killed 50 million people. HIV currently kills nearly 3 million people annually. If we add all other infectious diseases, we get a total of nearly 11 million deaths per year. (2006, 90)

> *Accidents*
> [There are] approximately 3.5 million accidental deaths (including over a million road accident deaths a year). (Benatar, 2006, 90)

Indeed, "[w]hen all other deaths are added, a colossal sum of approximately 56.5 million people died in 2001. That is more than 107 people per minute" (Benatar, 2006, 90). Benatar now asks the reader to consider the more intentionally inflicted suffering:

> One authority estimates that before the twentieth century over 133 million people were killed in mass killings. According to this same author, the first 88 years of the twentieth century saw 170 million (and possibly as many as 360 million) people 'shot, beaten, tortured, knifed, burned, starved, frozen, crushed, or worked to death; buried alive, drowned,...[hanged], bombed, or killed in any other of the myriad ways governments have inflicted death on unarmed, helpless citizens and foreigners'. (Benatar, 2006, 91)

> According to the *World Report on Violence and Health*, there were 1.6 million conflict-related deaths in the sixteenth century, 6.1 million in the seventeenth century, 7 million in the eighteenth, 19.4 million in the nineteenth, and 109.7 million in that most bloody of centuries—the twentieth. (Benatar, 2006, 91)

> Nor does the suffering end there. Consider the number of people who are raped, assaulted, maimed, or murdered (by private citizens, rather than governments). About 40 million children are maltreated each year. More than 100 million currently living women and girls have been subjected to genital cutting. Then there is enslavement, unjust incarceration, shunning, betrayal, humiliation, and intimidation, not to mention oppression in its myriad forms. (Benatar, 2006, 91)

Finally, the suffering is so great that 815,000 took their own lives in 2000 (Benatar, 2006, 91). Part of the problem is that Pollyannaism suggests that this suffering can be avoided. But some of what is listed above simply cannot be avoided. Likewise, even if some lives go better than Benatar suggests, "those (relatively) high-quality lives are exceedingly uncommon" (2006, 92). For any one of these 'charmed' lives there are literally millions upon millions of lives of poor quality. This means that "[e]ven the most privileged people could give birth to a child that will suffer unbearably, be raped, assaulted, or be murdered brutally. The optimist surely bears the burden of justifying this procreational Russian roulette" (Benatar, 2006, 92). Since there is no advantage to existing over not existing, Benatar fails to see how the risk could be justifiably imposed (Benatar, 2006, 92).[23]

Objections and Replies

Though it has received less attention than the Asymmetry Argument, there have still been a number of responses to the Quality of Life Argument in the philosophical literature. I now highlight some of the main objections and replies.

One objection to the Quality of Life Argument has to do with challenging Benatar's use of psychological evidence to support the claim that humans exhibit biases toward optimism (2012). Specifically, Brooke Trisel has suggested that studies on Pollyannaism suffer from selection bias because they focus on Americans who are college aged (2012).[24] Though it is true that some social science suffers from selection bias, particularly when it is conducted primarily on university campuses, the psychological literature that Benatar appeals to does not suffer from these problems (Benatar, 2012, 63). Indeed, some of the very papers appealed to by Benatar that Trisel criticizes explicit say that their work is repeatable in noncollege populations (see Taylor & Brown, 1988). Benatar notes that Matlin and Stang explicitly suggest many studies about optimism apply to populations beyond college students (2012, 147; see Matlin & Stang, 1978). The studies originally cited by Benatar include a wide variety of different subjects (2012, 147). Benatar explains that "although psychologists in general have focused disproportionately

[23] Notice that the statistics cited by Benatar are now around 20 years old. But this likely only makes matters worse because (i) the world's population has grown since then which means there is more suffering; and (ii) it is possible the average quality of life is actually worse than it was 20 years ago, though this is ultimately an empirical matter.

[24] Trisel cites Smart (1996) and Arnett (2008).

on undergraduates (and especially undergraduate psychology students) and on Americans, there is plenty of evidence that the optimism bias is characteristic of humanity more generally" (2012, 148).[25]

One common objection to the Quality of Life Argument is that pain can be instrumentally valuable (e.g., Trisel, 2012). For instance, part of the problem with leprosy is that it creates an inability for sensory nerves to feel pain. This means that people with leprosy are unable to protect their skin from damage like everyone else. Pain is often protective; it prompts humans to change their behaviour. Rivka Weinberg appeals to the example of Viktor Frankl to make a similar point, noting that he did not regret his suffering because it had enabled him to understand the meaning of life (2012). However, even if some pains are instrumentally valuable, it does not follow that they are not *intrinsically* bad (Benatar, 2012, 150). It also does not follow that some pains are far more severe than is necessary in order to be useful. Finally, though it is clear that some pain is indeed useful, Benatar writes that this fact is a bad-making feature of human life. In other words, "[i]t would be so much better if some pains were not necessary – if we could have the benefits that pain sometimes brings but to have those benefits without the pains" (Benatar, 2012, 150).

Another objection to the Quality of Life Argument says that the standards Benatar employs for assessing one's quality of life are inappropriately high. For example, Trisel argues that the standards for measuring the quality of a life "must be based on what is nomologically possible, not on logically possible, but nomologically impossible conditions" (Trisel quoted in Benatar, 2012, 150). Benatar responds that he sees "no reason why we should not judge the quality of life by the highest (logically) possible standard. There is nothing arbitrary about choosing perfection over any standard that approximates but does not reach it" (2012, 150). Benatar employs the following analogy by appealing to evaluating the work of students. In such a case, "we might employ a kind of perfection standard (100%) in judging the student's work. This does not mean that

[25] Benatar also points out that Trisel says nothing of the other two psychological features he discusses such as the fact humans tend to overrate their abilities and naturally adapt to more difficult circumstances (2012, 148).

we cannot use one threshold to determine a Pass and another to determine a First Class Pass. Setting a perfection standard is one thing, and benchmarking against that standard is another" (Benatar, 2012, 151).

Related to the above concern that it is inappropriate for Benatar to appeal to an objective perspective (i.e., what Benatar refers to as *sub specie aeternitatis*) when evaluating the quality of human life (Benatar, 2012, 152).[26] First, in light of the fact that humans arose from evolution, Metz believes that it is very unlikely human judgments "would be informed by the point of view of the universe" Metz 2011 252 quoted in Benatar, 2012, 152). Second, Metz believes that there is a potential inconsistency in Benatar because he offers "judgments of what is immoral and what is harmful… without appealing to the point of the view of the universe" (Metz, 2011, 253 quoted in Benatar, 2012, 152).[27] Benatar believes that these arguments "seem to be based on a mistaken interpretation of what it means to judge the quality of life *sub specie aeternitatis*" (Benatar, 2012, 151). The objective perspective need not be removed from human evaluation. The state of the 'human condition' limits our ability to imagine just how much better things could be compared to the current human experience (Benatar, 2012, 151). This means that "[t]he more objective perspective is required to show up these coping mechanisms for what they are and to provide a more reliable view of the quality of human life" (Benatar, 2012, 151–152).[28]

Prima Facie Consistency with Christianity

There is nothing in the Quality of Life Argument that is obviously inconsistent with Christianity. The most immediate objection may well be that Christianity is supposed to make the level of pessimism about the human condition in this argument inappropriate. This has less to do with challenging the quantity and quality of suffering described by Benatar

[26] Rivka Weinberg in Benatar (2012, 152) raises a similar concern.

[27] See also DeGrazia (2010, 327–32) and Hauskeller (2022) for additional challenges to objectivity. For a reply, see Benatar (2022, 123–127).

[28] DeGrazia argues that many people have good lives according to hedonism or desire satisfaction theories of a meaningful life (2010, 325–326). See Benatar (2010, 142–143) for a reply.

by this argument but is instead about the *attitude* that Christians are supposed to exhibit in the face of suffering. Indeed, one might suggest that Christians are to exemplify *joy* (see Galatians 5:22; 1 Thessalonians 1:6). However, it does not follow that because Christians are supposed to face their existence with a certain attitude, that procreation is permissible. Or at least this cannot be assumed, and significant argumentation for this conclusion would have to be given. Likewise, the Biblical data portrays many people of God sad and sorrowful (e.g., Job and Jeremiah). Jesus is described as a man of many sorrows (Isiah 53:3). Finally, Christianity affirms that the world in its current state is fallen and broken. It might be *finally* good, but in its current form it is not good.[29] Though there is more to say about how Christians should react to the Quality of Life Argument, there is little reason to think that Christianity is *prima facie* inconsistent with it.

3.4 The Misanthropic Argument

There are two main ways of categorizing arguments for anti-natalism that I have not mentioned until now. Philanthropic arguments for anti-natalism claim that it is impermissible to procreate given the harm individuals who come into existence in our world will inevitably face. The Asymmetry Argument and Quality of Life Arguments are philanthropic arguments. Misanthropic arguments, on the other hand, claim that procreation is impermissible given the harms that those who are procreated will likely cause other humans, non-human animals, and the environment. Benatar offers this latter type of argument in his chapter, "The Misanthropic Argument for Anti-natalism" (2015).[30]

[29] This is especially so for those within the Christian tradition that affirm the doctrines of original sin and original guilt.

[30] See Lougheed (2022b, 28–30) for discussion of the modal scope of this argument (see also Benatar 2022, 127–128).

The Argument

Here is Benatar's Misanthropic Argument for Anti-Natalism:

> (5) We have a (presumptive) duty to desist from bringing into existence new members of species that cause (and will likely continue to cause) vast amounts of pain, suffering and death.
> (6) Humans cause vast amounts of pain, suffering and death.
> Therefore,
> (7) We have a (presumptive) duty to desist from bringing new humans into existence. (Benatar, 2015, 35)

In what follows I will take Benatar's defense of each premise in turn.

In order to defend premise (6), Benatar appeals to human's generally poor impulses, their destructiveness towards one another, the suffering they cause other animals, and the damage that they do to the environment.

Harm to Humans

Benatar observes that throughout history and into the present-day, humans continue to have a rather high view of themselves (2015, 36). The term *homo sapiens* literally translates to 'the thinking human'. Benatar acknowledges that humans appear to think more than non-human animals and that we have 'greater technical capacity' than them (Benatar, 2015, 36). However, he also observes that though a very small percentage of elite humans have achieved remarkable accomplishments, "we would do well to note that these achievements are remarkable only because (1) they are not within the reach of most humans; (2) the cleverest humans typically produce them only by pushing the limits of their capacity; and (3) there are no more cognitively capable species on the planet to put our achievements into a humbling perspective" (Benatar, 2015, 36).

According to Benatar, then, it is unfair to judge the value of the human species on the basis of what so few humans can actually accomplish. Furthermore, it's not as if the elite also do not suffer from many of the other deficiencies that Benatar claims plagues humanity. For example, humans suffer from numerous cognitive biases. Intuitive or quick judgments are often mistaken, but we fail to take the time and energy needed to slow down and reflect. The 'framing effect' shows that our decisions can be influenced based on the way information is presented, even if that information is the same across cases. The 'endowment effect' demonstrates that people place irrationally high value on retaining something. Pollyannaism shows that humans are unduly optimistic. We also consistently exhibit poor impulse control in general (Benatar, 2015, 36). Consider that:

> For all the thinking that we do we are actually an amazingly stupid species. There is much evidence of this stupidity. It is to be found in those who start smoking cigarettes (despite all that is known about their dangers and their addictive content) and in the excessive consumption of alcohol – especially in those who drive while under its influence. It is to be found in the achievements of the advertising industry, which bear ample testament to the credulity of humanity. It is also to be found in the success of political sloganeering, demagoguery, and spin, to which billions of people fall prey. The seriousness with which so many people take matters of utter inconsequence – such as sport and the vicissitudes of particular sport teams – and the popular adulation of shallow, dysfunctional sports, music, and film stars are also items of evidence. Further signs are to be found in the fads and fashions and delusional obsessions that run rampant. (Benatar, 2015, 36–37)

With respect to premise (6), the problem is that these cognitive failings often cause, at least in part, humans to harm each other. We exhibit an extreme tendency towards conformity and following authority, even when doing so leads us to hurt each other (Benatar, 2015, 37). Studies have shown that humans have a high propensity to conform to the majority's answer, even when it is clearly wrong.[31] The willingness to follow authority has also led humans to do awful things to each other.

[31] Solomon (1956) is considered to be a ground-breaking study on conformity. See also Bond and Smith (1996). These and remaining additional sources from this chapter are all taken from Benatar (2015).

While many doubt that they would be the ones who would perpetrate unjustified harm by following authorities, studies suggest that most people would in fact follow authority (Benatar, 2015, 37–38).[32]

It is difficult to defend the claim that our intelligence compensates for our moral deficiencies in light of our destructiveness:

> Many hundreds of millions have been murdered in mass killings. In the twentieth century, the genocides include those against the Herero in German South-West Africa; the Armenians in Turkey; the Jews, Roma, and Sinti in Germany and Nazi-occupied Europe; the Tutsis in Rwanda; and Bosnian Muslims in the former Yugoslavia. Other twentieth-century mass killings were those perpetrated by Mao Zedong, Joseph Stalin, and Pol Pot and his Khmer Rouge. But these mass killings were by no means the first. Genghis Khan, for example, was responsible for killing 11.1% of all human inhabitants of earth during his reign in the thirteenth century [...] The gargantuan numbers should not obscure the gruesome details of the how these deaths are inflicted and the sorts of suffering the victims endure on their way to death. Humans kill other humans by hacking, knifing, hanging, bludgeoning, decapitating, shooting, starving, freezing, suffocating, drowning, crushing, gassing, poisoning, and bombing them. (Benatar, 2015, 39)

There seems hardly any end to the gruesome examples reported by Benatar. The Lord's Resistance Army, among others, kidnap children every year in order to make them child soldiers. Benatar explains that these children are sometimes:

> [F]orced to kill members of their own family or others, typically in gruesome ways. In one case a boy was told to pound to death the baby of a woman he knew. If the inductees refuse orders they are beaten savagely or even killed [...] It is estimated that there are currently about three hundred thousand child soldiers in conflicts in Asia, Africa, the Americas, and elsewhere. (2015, 40–41)[33]

---

32 See Milgram (1974) and Zimbardo (2007).

33 See Boothby and Knudsen (2000).

Contrary to what some believe, slavery is still practiced today, including sexual slavery.[34] Even if only a minority of people participate in such practices, many indirectly support them. For example, "[s]ometimes large numbers of people endorse a world-view in which 'honor killings' thrive, or in which terrorists are hailed as heroes. Sometimes the facilitation of evil results not from an endorsement of the evil but from stupidity, gullibility, dogma, or some other failing" (Benatar, 2015, 42). Many ordinary people can fall into mob violence, which has occurred frequently throughout human history. This includes not only lynch mobs but also consider that "a Wal-Mart employee in Valley Stream, New York, was trampled to death by shoppers who stormed into the store looking for bargains" (Benatar, 2015, 42).[35]

Humans also don't just murder each other. They "rape, assault, flog, maim, brand, kidnap, enslave, torture, and torment other humans" (Benatar, 2015, 40). Though these are the worst harms, humans also frequently "lie, steal, cheat, speak hurtfully, break confidences and promises, violate privacy, and act ungratefully, inconsiderately, duplicitously, impatiently, and unfaithfully" (Benatar, 2015, 43). Even if justice is sought, it is hardly ever achieved. Many of the evilest leaders in human history ruled for the course of their natural lives, while others had peaceful retirements or were only exiled (Benatar, 2015, 43). In sum, "'Bad guys' regularly 'finish first'. They lack the scruples that provide an inner restraint, and the external restraints are either absent or inadequate" (Benatar, 2015, 43).

Harm to Animals and the Environment

According to Benatar, the amount of suffering that humans inflict on animals each year is hard to fathom. Given that the vast majority of humans are not vegetarians or vegans, most of them are complicit in this suffering. Consider that ""[o]ver 63 billion sheep, pigs, cattle, horses, goats, camels, buffalo, rabbits, chickens, ducks, geese, turkeys, and other

---

34 See Polgreen (2008) and Kristof (2009).

35 See McFadden and Macropoulos (2008).

such animals are slaughtered every year for human consumption. In addition, approximately 103.6 billion aquatic animals are killed for human consumption and non-food uses" (Benatar, 2015, 44). These numbers exclude the hundreds of millions of male chicks killed every year because they cannot produce eggs. It also excludes the millions of dogs and cats that are eaten in Asia every year (Benatar, 2015, 44). Each year there are also 5 billion bycatch sea animals which are those caught in nets, but not wanted. Furthermore, "[t]he deaths of the overwhelming majority of these animals are painful and stressful" (Benatar, 2015, 44). The average meat eater will consume at least 1690 animals in their lifetime (a rather low estimate) which amounts to an extremely large amount of harm (Benatar, 2015, 54–55).[36]

Benatar also observes that the harm inflicted on animals extends far beyond using them for food. It seems that at least 115 million animals are experimented on every year (Benatar, 2015, 45).[37] While many of these experiments are no longer allowed, humans can still inflict severe pain on animals with tests. For example, mice have been genetically engineered to developed motor neuron degeneration and even cancer (Benatar, 2015, 46).[38] Still other animals are bred to race, fight, and entertain us. They often live in poor conditions and are euthanized when no longer able to perform (Benatar, 2015, 46). Cruelty even to domestic household pets remains rampant.[39] Consider that "[m]illions of dogs and cats are abandoned each year. In the shelters to which they are sent, the overwhelming majority are killed because homes cannot be found for them. It is astounding that in the context of so many unwanted domestic animals, humans actively breed more such animals" (Benatar, 2015, 47). Additionally, numerous animals are killed every year for their fur. Benatar observes that "[s]ome humans seem to think that fashion is a good reason to make an animal suffer and die" (Benatar, 2015, 48).

Finally, Benatar explains that humans are also incredibly destructive to the environment. The human population is growing exponentially

---

[36] See Appleby et al. (2004) and Braithwaite (2010).

[37] See Taylor et al. (2008).

[38] See Gurney et al. (1994) and Anderson (1988).

[39] See pet-abuse.com/cases.

and the negative environmental effects per person continues to increase too. This is partly due to industrialization and a steady growth in per capita consumption (Benatar, 2015, 48). As a result, there are "unprecedented levels of pollution. Filth is spewed in massive quantities into the air, rivers, lakes, and oceans, with obvious effects on those humans and animals who breathe the air, live in or near the water, or who get their water from those sources. The carbon dioxide emissions are having a 'greenhouse effect,' leading to global warming" (Benatar, 2015, 48).[40] CO_2 emissions per year per person are massive. While emissions are lower in developing countries, they tend to have much higher birthrates than those in wealthier countries. As the population increases, adding more humans will invariably harm the environment. All of these facts contribute to a cumulative case for the vast harm cause by humans.

Premise (5) is the normative premise of Benatar's argument. Notice that it does not claim that we should kill members of a dangerous species or prevent that dangerous species from procreating. Instead, it merely says "that one should oneself desist from bringing such beings into existence" (Benatar 2015, 49). In order for premise (1) to be true it also does not have to be the case that *every single member* of the species is dangerous. The likelihood that a new member of the species will cause significant harm is enough to make procreation too dangerous to be justified. Also notice that this premise is silent about the species in question. Benatar insists that it would be easily accepted if it were about some non-human species: "Imagine, for example, that some people bred a species of non-human animal that was as destructive (to humans and other animals) as humans actually are. There would be widespread condemnation of those who bred these animals" (Benatar, 2015, 49).

One might worry that "there is something paradoxical about claiming that we have a duty to desist from bringing into existence members of a species that is harmful to itself. There is, on this view, something odd about citing the harm caused to humans by humans as a reason to desist from creating humans" (Benatar, 2015, 49). If humans are valuable enough to be protected in this way, then they cannot really be

[40] Benatar also notes that even for climate sceptics there is still more than enough support for premise (2).

the awful species that Benatar describes in defending premise (2). So, the misanthropic argument appears to be in tension with philanthropic arguments for anti-natalism. For recall that philanthropic arguments for anti-natalism focus on the harm that will inevitably befall the humans who are brought into existence.

Benatar offers a number of responses to this worry. He suggests that "it is a mistake to muddle our attitudes towards victims and our attitudes to perpetrators – even when the victims are also perpetrators" (Benatar, 2015, 50). We tend to think there are limits with respect to punishment because even perpetrators matter morally. Consider that we do not rape and torture perpetrators of such crimes as punishment. For "[a] a woman may be guilty of physically assaulting her child, but that does not mean that we should be unconcerned about the physical assault her husband inflicts on her, or that we should not be concerned about the violence he suffers at the hands of others" (Benatar, 2015, 50).

The Presumptive Duty Not to Procreate

Presumptive duties are defeasible. Such a duty only holds if there are no good reasons to do otherwise (Benatar, 2015, 51). So one possible way of avoiding the conclusion of the misanthropic argument is to counter that the good humans do is pervasive enough to often defeat this presumptive duty. If this is right, then procreation will often be permissible (Benatar, 2015, 51). However, in light of the vast harms that humans cause, meeting the burden of proof regarding the amount of counteracting good humans are responsible for, is going to be extremely difficult. Benatar believes that there are no reasonable prospects "of reducing human destructiveness to negligible levels fairly promptly and then ensuring that they do not rise again" (Benatar, 2015, 50). If there were such prospects, it might be reasonable to focus on reducing human destructiveness instead of ceasing to procreate. But Benatar believes that "we cannot expect that human destructiveness will *ever* be reduced to such levels. Human nature is too frail and the circumstances that bring out the worst in humans are too pervasive and are likely to remain so" (Benatar, 2015, 51). Thus, he concludes that it is "naïve utopianism to think that a species as destructive as ours will cease, or all but cease, to be destructive" Benatar, 2015, 51).

There has been comparatively little interaction in the literature with this argument compared to the Asymmetry Argument and the Quality of Life Argument. Perhaps part of the reason for this is that some of the objections leveled at the other two arguments are also thought to apply to it. Likewise, this chapter was publishing years after *Better Never to Have Been* and so may perhaps garner more responses in the future. It is difficult to see how the empirical evidence Benatar appeals to could be disputed to the extent that it would be reasonable to deny that (almost all) humans cause significant harms in their lives. The crux of the debate about this argument will be about what can be inferred from that fact.

Prima Facie Consistency with Christianity

There is nothing in the Misanthropic Argument that is obviously inconsistent with Christianity. In fact, it is perhaps the argument that most intuitively coheres with at least some aspects of the Christian tradition. Certain strands of Christianity emphasize the concepts of original sin and original guilt. The former says that humans born into a fallen world have a propensity to sin where presumably much sin involves causing harm to others. The latter says that humans born into a fallen world are guilty in light of Adam's sin and therefore born in need of redemption.

3.5 Anti-Natalism, Pro-Mortalism, and Speciecide

The Objection

To conclude this chapter, a general objection to anti-natalism must be addressed since it is sometimes brought against Benatar in various forms. The objection, if successful, is supposed to show that the entailments of anti-natalism are absurd and as such that view should be rejected. Though it can perhaps be phrased formally as a *reductio ad absurdum* of anti-natalism, I will not focus on explicitly explaining the objection in those terms. Instead, I note two different ways that the objection can be formulated. First, there is the worry that anti-natalism entails pro-mortalism such that it recommends suicide for all or most individuals. Of course, so the objection goes, any plausible ethical view cannot recommend suicide. Second, a similar worry is that an implication of anti-natalism is that the human species should be exterminated. Again, this is supposed to indicate that anti-natalism is absurd.

Possible Replies

By way of initial response, Benatar has been crystal clear throughout all of his writings that anti-natalism does not recommend suicide. This is because there is a difference in evaluating whether a life is worth starting versus whether a life is worth continuing. For Benatar, it is never the case that a life is worth starting, though he affirms that most of us have an interest in continuing to live. Perhaps most importantly, Benatar claims that "it is possible to think both that coming into existence is a serious harm and that death is (usually) a serious harm. Indeed, some people might think that coming into existence is a serious harm in part because the harm of death is then inevitable" (Benatar, 2013, 148).

With respect to the fact that anti-natalism, if adopted, should prompt speciecide is simply mistaken. As stated in 1.4, it could be that there is no way to avoid violating rights in order to enforce anti-natalism. For example, Benatar says that though he thinks "that procreation harms those brought into existence and that there is thus no moral right to procreate, there ought nonetheless still [to] be a legal right to reproduce. The arguments for this apply, a fortiori, to a legal right not to be killed, thus ruling out speciecide" (Benatar, 2013, 149). Likewise, there is reasonable disagreement about anti-natalism such that anti-natalist policies should not be enforced (see Benatar, 2006, 106–113; 2013, 149).

If the objection instead is not that speciecide would occur, but that the implication of human extinction shows the view is absurd, this is again simply mistaken. The vast amounts of suffering associated with human existence would cease if humans went extinct and this is a good thing. Such a great reduction in suffering cannot be thought to show the viewing entailing that reduction is absurd. The objector in this case needs to engage the actual arguments offered by Benatar.

Extinction as the Highest Priority?

Joseph Packer claims that Bentar's Asymmetry Argument "elevates extinction to the highest priority and thus justifies violence against people that currently exist in order to achieve it regardless of whether Benatar personally endorses involuntary extinction" (2011, 228). Furthermore, Packer claims that "[e]xtinction would prevent a potentially infinite number of future humans, thus it prevents a potentially infinite amount of future suffering with no pleasure to offset it. The magnitude of infinite future suffering is so great that no amount of finite suffering to achieve extinction could be worse" (2011, 228). Accordingly, given that voluntary extinction is very unlikely, 'coercive efforts' to reach extinction are justified (Packer, 2011, 229).

Benatar affirms that extinction would be good but denies that it is the very highest priority such that it should be pursued at any cost (2013, 149). Likewise, "One problem here is that bringing about extinction by killing does not prevent an infinite amount of future suffering. This is because extinction will later occur whether or not we aim to bring it about. Thus speciecide will prevent an enormous but finite amount of suffering" (Benatar, 2013, 149). Packer also doesn't acknowledge the incredibly high likelihood that speciecide would also fail, in addition to the great amount of harm such attempts would cause (2013, 149). Benatar also notes that human rights plausibly constrain the type of policies we can enact such that an anti-natalist program is impermissible no matter the potential benefits associated with enacting it (2013, 149).

Now, Packer claims that even if an appeal to rights can avoid the entailment of speciecide, "it should still leave one eager for an involuntary world-ending event like a major asteroid collision. An asteroid collision could cause human extinction and thus reduce the total amount of suffering" (2011, 230). To press the point further he writes:

> Assume a car accident kills a young pregnant woman and her husband. Absent some bizarre circumstances most people would say that this was a tragedy. Given Benatar's logic, however, one may be compelled to view the accident as a blessing. The loss of the couple's lives (lives that would be filled with pain as well as pleasure) needs to be weighed against all of the pain that the unborn child would have suffered over her or his life. Remember also that if the young woman gave birth to the child it too would at some time die. I do not claim to be able to quantify the value of life precisely, but one could imagine many scenarios where given Benatar's asymmetry we should view the car crash as better than the birth of the child. (2011, 230)

But this is so much worse than dying of natural causes. Such an accident would cause much pain and suffering for the family and friends of the victims. Imagine the first responders to the crash who have to witness the carnage. As Benatar says, death is a great harm and one that most people reasonably seek to avoid. In an attempt to push this objection one step further, Packer writes the accident results in much good because none of the victims will have any (more) progeny:

> Benatar points out that if the average couple has three children then in ten generations each couple will have 88,572 descendents (Benatar 2006, pp. 6–7). The combined suffering of 88,572 people almost certainly outweighs the lives of even the happiest two progenitors on the planet today. (Packer, 2011, 230)

Again, there is nothing inconsistent in Benatar claiming that though not having progeny would be good, the harm involved in the accident is still something to be avoided. Even a catastrophic event that caused the extinction of humanity all at once, would set back my interests such that I am rational to think of such an event as bad.

Finally, notice that Benatar's (i) to (iv) procreative asymmetries are intuitively obvious as is the Basic Asymmetry (albeit to a lesser extent). The facts about the suffering of humans (i.e., the Quality of Life Argument) and the harm humans tend to cause (i.e., the Misanthropic Argument) are difficult to deny. It is intuitively obvious that one both experiences and causes much harm throughout their life. Notice, therefore, that it is *not* the premises or initial starting points of anti-natalism that are counterintuitive. Instead, it is the conclusion of anti-natalism and what it (allegedly) entails that is the problem. So, Packer's objection prioritises intuitions at the level of conclusions/implications and seemingly ignores them at the level of premises. There may well be good reasons for this methodological choice, but an argument needs to be given. Otherwise, it seems that Packer has failed to address Benatar's reasons for anti-natalism.

Is Life So Bad We Should End It Immediately?

It might be thought that if one really internalised just how bad life really is according to Benatar, most people would be inclined to commit suicide (see Harman, 2009). Perhaps anti-natalism leaves the question of suicide as a individual choice where choosing to commit suicide is always a rational action. But there are at least two problems with this objection. First, Benatar claims that for many people, the worst parts of their lives do not occur in the first part of life, thereby making it reasonable to delay suicide. Second, since death is a great harm, it is reasonable to think that one's life has to become particularly bad for suicide to be a reasonable response to suffering (Benatar, 2013, 150).[41]

3.6 Conclusion

This chapter could be much longer given that Benatar's work on anti-natalism continues to generate much interaction in the literature. The purpose of this chapter was to give a detailed explanation of each of his three main arguments for anti-natalism, along with exploring some of

---

[41] For a few more objections and replies on this issue (see Benatar, 2012, 157–160; Benatar, 2022, 143–148; McGregor & Sullivan-Bissett, 2012; Sullivan-Bissett, 2022). Torres (2020) argues that it is possible to be an anti-natalist while simultaneously hoping that humans continue to survive. He argues this position is possible if life-extension technologies allowed humans who already exist to continue to live indefinitely. Thus, humanity could survive even if it does not reproduce.

the objections to those arguments. Given the vast literature, I made no attempt at completeness, but hope to have provided a bird's eye overview. Instead, my main purpose was to determine which of Benatar's arguments, if any, are at least *prima facie* compatible with Christianity. I argued that the Asymmetry Argument, the Quality of Life Argument, and the Misanthropic Argument are all compatible with Christianity, or at least not obviously at odds with it. In the next chapter I will outline other arguments for anti-natalism in the literature, again exploring whether they are *prima facie* consistent with Christianity.

4

Additional Arguments for Anti-Natalism

4.1 Introduction

This chapter focuses on outlining arguments for anti-natalism in the philosophical literature beyond the ones that have been developed by David Benatar. As with the arguments from Benatar, I will also examine the extent to which these arguments are *prima facie* compatible with Christianity and so may be appealed to in the next chapter. The degree to which these arguments have been developed varies greatly, and as such my main aim is to provide a bird's eye view of what I take to be the most important components of them. In what follows I outline the arguments based on Risk (4.2) Hypothetical Consent (4.3), Exploitation (4.4), not creating Victims (4.5), failing to guarantee the Right to Security (4.6), and Negative Utilitarianism (4.7). I argue that all but the argument based on negative utilitarianism are at least *prima facie* consistent with Christianity (4.8). I conclude by observing that a common theme across many of these arguments focuses on the imposition of the risks associated with coming into existence and the consent (or lack thereof) to such risk.

K. Lougheed, *Christian Anti-Natalism*, https://doi.org/10.1007/978-3-032-03902-6_4

4.2 The Risk-Based Argument

There are a variety of ways in which a risk-based argument for anti-natalism could be developed. In what follows I focus on a recent analysis of such arguments by Erik Magnusson. He develops risk-based arguments based on brief comments from Benatar (see Benatar, 2006, 92; Magnusson, 2022, 103). For Benatar, the Quality of Life Argument (see 3.3) does not have to entail that every person brought into existence experiences significant harm; rather, it just needs to be the case that there is a significant *risk* of experiencing such harms.[1]

Risk based arguments need to establish the conditions under which it is wrong to impose risk, how bringing a child into existence meets those conditions, and then establish whether procreation is all-things-considered permissible in such a case. This means that risk-based arguments for anti-natalism should be structured as follows:

> (1) It is impermissible to impose a risk of harm on others under *x* conditions;
> (2) Bringing a child into existence involves imposing a risk of harm on the child in a way that satisfies *x* conditions.
> Therefore,
> (3) It is impermissible to bring children into existence. (Magnusson 2022, 104)

(1) can be challenged in a number of different ways. Assessing whether imposing a risk is permissible involves figuring out how serious the potential harms are, the likelihood they will occur, and the justification for the imposition in the first place (Magnusson, 2022, 104). (2) can be challenged by attempting to show that procreation does not involve imposing the risks in question or that the risks of the harm are much lower than supposed (Magnusson, 2022, 104–105).

Magnusson also writes that:

---

[1] Magnusson believes it is worthwhile to develop a risk-based argument because it is less controversial than Benatar's other arguments (2022, 103). His reason for this is that the risk assessments used to establish the argument are often appealed to other less controversial contexts. They also do not have to say the harms of being brought into existence in inevitable.

> [R]isk-based arguments must by definition focus on *atypical* harms, or harms of a type or magnitude that human beings will not necessarily experience over the course of their lifetimes—otherwise, the argument simply reverts to a harm-based argument, or one intended to show that bringing a child into existence is impermissible in virtue of causing that child to be harmed. (2022, 105)

Now, by *atypical* I think Magnusson just means *not guaranteed*. Otherwise, it is unclear what he could mean by it. Cancer is therefore atypical in this sense. As a Canadian male I have an approximately 45% chance of developing some type of cancer in my lifetime.[2] So, it is not atypical in that almost half of the men I know in Canada will develop cancer at some point in their lifetime. However, it is not guaranteed that any particular man in Canada will develop it. Drawing a distinction between inevitable harm and the risk of harm shows how risk-based arguments differ from the Asymmetry Argument and Quality of Life Argument where those arguments rely on claims about actual harm experienced. Risk-based arguments do not rely on the claim that the harms in question become actual.[3]

Magnusson offers three different types of risk-based arguments that fit the suggested structure of any argument for anti-natalism based on risk. He begins with Benatar's versions, before moving to a stricter version (i.e., extremely risk adverse), and closing with his preferred risk-based argument which focuses on the *essential interests* of children who will have to bear the risks imposed on them by their procreators. However, in the end he appears to cast doubt on whether that version should be accepted.

---

2 Lifetime probability of developing and dying from cancer in Canada - Statistics Canada (statcan.gc.ca).

3 For more on 'pure' risk-taking see Thomson (1986) and Oberdiek (2017, ch. 3).

Magnusson on Benatar's Risk-Based Argument

Magnusson begins by reconstructing a risk-based argument from Benatar before showing how it could be improved upon. He claims that Benatar's argument assumes the following principle:

> *Risk Principle #1:* it is impermissible to non-consensually impose a risk of catastrophic harm on others when there is a high probability of occurrence. (Magnusson, 2022, 106)

The harms in question here are one's described by Benatar in his Quality of Life Argument (see 3.3). These are harms such that if they were to occur, they could reasonably make a person's life not worth living. When entered into the formula for risk-based arguments, Risk Principle #1 shows:

> (4) It is impermissible to non-consensually impose a risk of catastrophic harm on others when there is a high probability of occurrence.
> (5) Bringing a child into existence involves non-consensually imposing a highly probable risk of catastrophic harm on that child.
> Therefore,
> (6) It is impermissible to bring children into existence. (Magnusson, 2022, 106; *numbering mine*)

Though Magnusson acknowledges there is 'inherent risk' involved in procreation, he says that Benatar's "analogy with Russian roulette is surely unwarranted" (2022, 108). He continues:

> While firing a loaded pistol at a child presents a very high probability of a catastrophic outcome—particularly, say, if one in six chambers are loaded—the same cannot be said in good faith about procreation. We can never predict the future, but we can normally make reasonable projections about the kind of lives our children will lead based on the range of information that is available to us, including the average quality of life in the society we live in, the risks that are associated with living in that particular society, and our personal capacity to mitigate those risks and provide for our children's emotional and material well-being. In many cases, it is

> reasonable to conclude that the risk of the relevant catastrophe—a life so immersed in suffering as to be of questionable value to the person who leads it—is sufficiently low to render our decision to procreate justifiable on the terms of *Risk Principle #1*. (Magnusson, 2022, 108)

Contra Benatar, listing the bad things in life does not tell us how likely they are to happen (Magnusson, 2022, 109). Though Magnusson says that:

> Benatar might attempt to undermine it in a different way by establishing that the probability of catastrophic harm is actually much higher than we tend to realize. He pursues this strategy in the most recent statement of his risk-based argument, citing cancer as an example of a "very bad" fate that awaits many of us, with as many as one in two men and one in three women likely to develop it over the course of their lifetime. This strategy seems unlikely to succeed, however, for as terrible as it can be to suffer from a disease like cancer—or to witness that suffering in a loved one—it is debatable whether it falls into the category of catastrophic harms, or harms that would cause us to question whether a person experiencing them could have a worthwhile life. (Magnusson, 2022, 109)

In order to support this claim, Magnusson asks the reader to consider the following case:

> A cancer patient is suffering in the hospital when a group of doctors offer him an experimental drug. The drug is very likely to cure his particular type of cancer, though it comes with the side-effect of severe memory loss, such that it will cause him to forever lose memory of his family, friends, and other close relationships. If cancer were indeed a catastrophic harm in the sense required by Benatar's response, then it would be rational for the patient to accept the drug and cure himself of cancer at the expense of his relationships. (2022, 109)

Reframing Benatar's Risk-Based Argument

Part of the problem with Magnusson's objection here is that much of Benatar's point is that if the person did not exist, they would not be in the position with respect to choosing about different cancer treatments in the first place. The very fact of the person being in that position is evidenced they are harmed by being brought into existence. It is also extremely surprising that Magnusson does not think that cancer is an atypical harm (meaning we are not all guaranteed to get it) and as such that it is not a catastrophic harm. Not only is this bizarre; it is extremely callous. Experiencing a life-threatening cancer is surely a catastrophic harm. On Magnusson's view, it is difficult to understand what *could* count as a significant harm. He seems to think there has to be a high probability a severe harm will occur to make imposing the risk of that harm impermissible, but then he suggests that something as terrible and likely to happen as cancer does not count as one.

Benatar himself suggests that Magnusson has mischaracterized the nature of his gesture at a risk-based argument for anti-natalism. Benatar states that in order to accurately represent his views, Risk Principle #1 should be modified to:

> (4'). It is impermissible to non-consensually impose a risk of catastrophic harm on others when there is a high probability of occurrence, *and there is insufficient justification for the imposition of that risk of that harm.* (Benatar, 2022, 149)

This formulation means that Benatar is neutral with respect to whether the interests that are advanced must be that of the person on whom the risk is imposed, or if the interests of others could justify the imposition of risk. So, this also means revising (5) to:

> (5'). Bringing a child into existence involves non-consensually imposing a highly probable risk of catastrophic harm on that child, *with insufficient justification for imposition of that risk of that harm.* (Benatar, 2022, 149)

Benatar also rejects Magnusson's notion of catastrophic harm, which is that the harm experienced could make a life not worth living (2022, 149). Consider that "[w]hether life is 'worthwhile' can connote a *purpose* rather than the *quality* of life. Similarly, life can have 'at least some value' even in the face of catastrophe. Indeed, sometimes a harm is catastrophic precisely because it threatens life" (Magnusson, 2022, 149). Benatar also believes "far too many people have very low standards for what counts as a life worth living. A harm does not have to meet those low standards in order to count as catastrophic" (2022, 150). This may explain why Magnusson fails to categorize cancer as a catastrophic harm. But a life-threatening cancer diagnose is certainly 'an event causing damage or suffering'.

Benatar also suggests that another "problem is that Dr Magnusson's interpretation of "catastrophic harm" is far more demanding than it needs to be, in order to accord with our ordinary standards about when the imposition of risks is unacceptable" (Benatar, 2022, 150). Ordinary standards suggest that there is an extremely strong default to refrain from exposing non-consenting individuals to a high likelihood of suffering, including from a condition such as cancer. This means there needs to be exceptional justification for posing such risks. Benatar suggests:

> [I]f smoking in the presence of a non-consenting smoker put the latter at an approximately 38.5% chance of developing lung cancer, I think it would be clear that smoking in the presence of a non-consenting non-smoker would be wrong. Indeed, many people think that smoking in the presence of non-consenting non-smokers is unacceptable even with the actual, much lower chances of causing them to have cancer. (2022, 150)

Benatar admits that a list of harms is not an argument, but notes that Magnusson only addresses cancer, not the many other harms that can and do befall some humans. Benatar argues that "[i]t is quite clear that if the risk of only cancer is as high as it is, the cumulative risk of all the possible catastrophes that can befall us is outrageously high" (2022, 150).

With respect to (5'), it may well be that exposing someone to a high risk of catastrophic harm is permissible in certain cases, but it is difficult to see how it could usually be the case (Benatar, 2022, 150). Benatar's statement allows that the justification *could* be based on the interests of others but says it is difficult to know which interests could ever justify procreation.[4] For example, it is true that phased extinction would be

[4] Häyry argues for anti-natalism on the basis that there are no child-regarding reasons to procreation and interests of prospective parents, etc., cannot justify it (2023).

very difficult for the last remaining people, but it is hard to see how this harm could outweigh the vast risk imposed in continuing to procreate. The interests of procreating or child rearing do not outweigh the risk of harms associated with existence (Benatar, 2022, 151). Consider:

> [A]n imaginary scenario in which a couple's procreating would put some *other* couple at a very high risk of suffering from cancer. It should be clear that in such circumstances, it would be wrong for the initial couple to procreate. If that is the case, why should procreation become acceptable in the actual cases in which the offspring (rather than some other couple) is put at high risk of cancer as a result of their parents' procreating? (Benatar, 2022, 151)

It is precisely for this sort of reason that the analogy between Russian roulette and procreation is justified. I will appeal to this type of Risk-Based argument in 5.2.

A Risk-Based Argument from Strictly Against Risk Imposition

The second type of risk-based argument for anti-natalism that Magnusson explores is one that is highly averse to imposing risk:

> *Risk Principle #2:* it is impermissible to non-consensually impose a risk of catastrophic harm on others regardless of its probability of occurrence. (Magnusson, 2022, 110)

When put into the formula for risk-based arguments, the following results:

> (7) It is impermissible to non-consensually impose a risk of catastrophic harm on
> others regardless of its probability of occurrence.
> (8) Bringing a child into existence involves non-consensually imposing at least some
> risk of catastrophic harm on that child.
> Therefore,

(9) It is impermissible to bring children into existence. (Magnusson, 2022, 110)

This style of argument can be located in the work of Matti Häyry (2004). He argues that it is impermissible to impose a risk of the worst outcomes on a person, no matter how small the risk (Magnusson, 2022, 111; see also Häyry, 2004, 377–378). Häyry appeals to Rawls' veil of ignorance to justify this claim. Suppose you do not know which life your child will get assigned. If they choose not to have a child, nobody is harmed or benefitted so its value is zero. But if they have a child, its life can be good or bad such that the value of the choice is negative, zero, or positive. It is best to avoid the possibility of a negative outcome when the guarantee of the value of zero (or a neutral one) is available (Magnusson, 2022, 111; see Häyry, 2004, 377–378).

Magnusson does not deny that some lives are not worth living but notes that in Rawls' original position, people are also shielded from probability estimates (2022, 111). Thus, with respect to procreation:

> Insofar as we can make reasonable projections about the kinds of lives our children are likely to lead based on the range of information that is available to us, we are not forced to assign equal probabilities to positive and negative outcomes. In fact, because a negative outcome in this context is defined as a life so low in quality as to make non-existence preferable, we can normally proceed confidently in the knowledge that the probability of this outcome is extremely low. (Magnusson, 2022, 112)

I think Magnusson here reflects a wildly optimistic view of the human experience. Again, consider that I have an almost 50% chance of getting cancer in my lifetime. Whether Magnusson's criticisms of this argument succeed ultimately may depend on the risks of having a bad life. However, in this context, the fundamental disagreement here between Magnusson and the anti-natalist seems to be not only about the odds of having a bad life but about what specifically constitutes a bad life.

The Essential-Interests Risk-Based Argument

The final risk-based argument for anti-natalism that Magnusson analyzes considers *essential interests.* He believes that one problem with the above risk-based arguments is that they tend to ignore the reasons for imposing the risk on a prospective child in the first place. But this matters because the magnitude of the harm, the likelihood it will happen, and the justification for imposing all matter when formulating risk-based arguments (Magnusson, 2022, 113). In light of this, Magnusson offers a principle inspired by Seana Shiffrin (see 4.3):

> *Risk Principle #3:* it is impermissible to non-consensually impose a risk of catastrophic harm on others *unless* doing so is necessary to advance their essential interests. (Magnusson, 2022, 114)

According to Magnusson, an interest is essential "when its satisfaction is a necessary precondition of living a minimally decent life and of pursuing other valuable projects and aims" (2022, 114). This includes:

> [O]ur interest in the continuation of our lives for the foreseeable future, in minimal physical health and rigor, in the integrity and normal functioning of one's body, in the absence of extreme pain and disfigurement, in basic subsistence and material security, and in freedom from unjustified interference and coercion, among others (Magnusson 2022, 114; see also Feinberg 1984, 37). An action can be said to advance someone's essential interests if failing to perform that act means the person is 'worse off' with respect to their essential interests. (Magnusson, 2022, 114)

This means that a high-risk, but life-saving surgery can be attempted because if it is not attempted the patient would be worse off regarding their essential interest in continuing their life. This is so even though there is a risk that the surgery will fail (Magnusson, 2022, 114).

Risk Principle #3 generates the following argument for anti-natalism:

> (10) It is impermissible to non-consensually impose a risk of catastrophic harm on others *unless* doing so is necessary to advance their essential interests;

> (11) Bringing a child into existence involves non-consensually imposing a risk of catastrophic harm on that child that is not necessary to advance their essential interests; therefore,
> (12) It is impermissible to bring children into existence. (Magnusson 2022, 114)

Magnusson suggests that an advantage of this argument is that it shifts focus from whether a risk will become an actual harm to about the reasons a person has for imposing the harm in the first place (Magnusson, 2022, 114). This means that even if the risk of a significant harm is incredibly low, there may not be justifying reasons for it. Magnusson asks the reader to:

> Imagine, for example, that I can press a button that presents a one-in-ten billion chance of delivering a major electrical shock to an unsuspecting victim. The chance of causing any harm by pressing this button is remote, though if there is no good reason for pressing it to begin with, this remote chance can still ground a strong moral presumption against doing so. (2022, 114)

Thus, even if the chances of a person experiencing a catastrophic harm are incredibly low in their life, there is no reason to impose that risk on a potential child since it is not in their interests (Magnusson, 2022, 115). It cannot be in their interests, since non-existent beings do not have any type of interests. If this is right, then procreating is impermissible.

Magnusson recognises that the main objection to this argument is going to be to (10) by suggesting that the essential interests of a third party might justify imposing the risk of catastrophic harm on someone else. Or perhaps it could be justified to advance significant non-essential interests (Magnusson, 2022, 115). Or maybe there are cases where the risk can be imposed to get a significant non-essential, non-comparative, benefit as evidenced in the following case:

> *Child Prodigy*: Aurora is a young girl who, from a very early age, displays an incredible talent for musical composition. Eager to encourage this talent, Aurora's parents enrol her into a rigorous training program

> throughout her childhood where she attends a special school for musically gifted children and spends nearly all of her free time in lessons and rehearsals. Eventually, Aurora grows up to be a happy and successful pianist who develops a novel playing style and pushes the boundaries of piano composition. (Magnusson, 2022, 116)

But this means that:

> By enrolling her into a rigorous training program throughout her childhood, Aurora's parents have exposed her to a risk of serious harm for the sake of delivering a significant non-essential benefit. There is a risk their decision will backfire, leaving Aurora miserable, isolated, and regretful that she did not get to experience a more carefree childhood, though there is also a chance their decision will yield a significant non-essential benefit, namely, the benefit of becoming a successful and prodigious musical talent. (Magnusson, 2022, 116)

This example shows that if enrolled in a rigorous training program, Aurora's essential interests will not be harmed. But to get the potential benefit she must be enrolled in the training as a child, before she can consent to it (Magnusson, 2022, 116). Magnusson argues that it must be shown why the example is wrong in order to establish anti-natalism.

The Revised Risk-Based Argument

There are a number of ways the anti-natalist could respond to Child Prodigy. First, it is disanalogous to the case of procreation. In the example, Aurora already exists and so can stand to gain a significant non-essential and non-comparative benefit. But a potential person does not exist and so cannot lose out on any sort of benefit. Now, this means Magnusson has to respond that the type of non-essential, non-comparative benefit that he has in mind is with respect to *third parties* even though his example suggests much of the benefit falls on Auora herself. So, he has to say that it is society at large that will benefit from Aurora's successful music career just as it is the parents who will benefit from having a child. This means that Risk Principle #3 has to be modified to:

> *Risk Principle #3'*: it is impermissible to non-consensually impose a risk of catastrophic harm on an individual *unless* doing so *possibly yields non-essential and non-comparative benefits on a third party.*

But this principle is false. Or more carefully, it requires further clarification that would cast doubt on whether procreation could be justified by it. Consider that the 'possibly' in it is vague. How likely do the benefits in question have to be to obtain? Though Magnusson presses on something important by trying to emphasize reasons instead of probability, the likelihood of catastrophic harm in comparison to gaining the non-essential and non-comparative benefits really matter. To see this consider a modified explanation of Child Prodigy:

> By enrolling her into a rigorous training program throughout her childhood, Aurora's parents have exposed her to a risk of serious harm for the sake of delivering a significant non-essential benefit. These harms include a significant chance of experiencing at least one of cancer, assault, poverty, HIV, war, famine, or other various diseases. There is a risk their decision will backfire, leaving Aurora to experience one or some of these catastrophic harms though there is also a chance their decision will yield a significant non-essential benefit, namely, the benefit of becoming a successful and prodigious musical talent. (Modified from Magnusson, 2022, 116)

This modification brings Child Prodigy much closer to procreation. Remember that the proponent of a risk-based argument does not say that these harms will befall a person, as the Quality of Life Argument seems to imply. But even if the risk of such significant harms are relatively low, it is doubtful whether they could be imposed based on the justification of musical excellence. Is it really permissible to impose the risk of cancer on someone in order to get the benefit of wonderful music? Perhaps the example could be constructed such that the probability of harm is so miniscule that imposing the risk is worth it. But then such a case would be even further away from procreation in the actual world. The odds of an individual experiencing significant harms in their lifetime is extremely high. To take just one example, even if a person does not get cancer, the odds of someone they love and care deeply about getting cancer is close to guaranteed.

I do not deny that the reasons for imposing the risk of a particular harm are important, but Magnusson cannot avoid discussion of the specific harms and the likelihood they will occur. Once they are considered, it is difficult to understand how Child Prodigy could justify procreation. Not only is it disanalogous to the case of procreation because Aurora already exists, but it does adequately consider the vast array of significant harms that are quite likely to come with existence. This means that Risk Principle #3' should be further modified in combination with Benatar's 1':

> *Risk Principle #3":* it is impermissible to non-consensually impose a risk of catastrophic harm on others when there is a *non-trivial chance* of occurrence, and there is insufficient justification for the imposition of that risk of that harm *which includes justification based on a non-trivial chance of advancing a third party's essential-interests or significant non-essential interests.*

Indeed, this modification may turn out to make the principle even stronger than Benatar's formulation because it is not the case that the chance of the significant harm occurring needs to be very high to make procreation impermissible. And even if Magnusson disagrees with this, it is still the case that the odds of significant harms occurring to someone who is born are extremely high. This is uncontestably so unless Magnusson wants to insist that something as terrible as terminal cancer is not a catastrophic harm. I suspect many natalists will find such an insistence unpalatable.

4.3 The Hypothetical Consent Argument

The next argument for anti-natalism that I examine is the Hypothetical Consent Argument. The basic idea of the argument is that procreation imposes unjustified harm on an individual because they did not consent to it (Harrison, 2012; Shiffrin, 1999; Singh, 2012). The basis of this argument is located in the work of Seana Shiffrin's discussion of bestowing *pure* benefits on individuals, though it is unclear whether she herself would fully embrace anti-natalism (Shiffrin, 1999).[5]

The Argument

Different cases help make clear why imposing certain types of harms are unjustified. To begin, consider a case where harming a non-consenting individual is clearly justified:

> *Rescue.* A man is trapped in a mangled car that apparently will explode within minutes. You alone can help. It appears that the only way of getting him out of the car will break his arm, but there is no time to discuss the matter. You pull him free, breaking his arm, and get him to safety before the car explodes. (DeGrazia, 2012, 151)

It is permissible in this case to impose the harm of breaking the man's arm without his consent because doing so prevents the greater harm of his death. In this case there is *hypothetical* consent to break the man's arm because he would have (or rationally should have) consented to the harm if he could. But now consider a different case that is also frequently discussed:

> *Gold manna.* An eccentric millionaire who lives on an island wants to give some money to inhabitants of a nearby island who are comfortably off but not rich. For various reasons, he cannot communicate with these

[5] Häyry and Sukenick suggest that Julio Cabrera drew attention to the 'one-sidedness' of procreation and is therefore a precursor to Shiffrin. Cabrera focuses on the risk of manipulating, harming others, and not being in control of the end of one's own life" (Häyry & Sukenick, 2024, 43–44; see Cabrera, 1996, 1989).

> islanders and has only one way of giving them money: by flying in his jet and dropping heavy gold cubes, each worth $1 million, near passers-by. He knows that doing so imposes a risk of injuring one or more of the islanders, a harm he would prefer to avoid. But the only place where he can drop the cubes is very crowded, making significant (but nonlethal and impermanent) injury highly likely. Figuring that anyone who is injured is nevertheless better off for having gained $1 million, he proceeds. An inhabitant of the island suffers a broken arm in receiving her gold manna. (DeGrazia, 2012, 151–152)

What makes the actions of the eccentric millionaire impermissible in this case is that the benefit imposed does not involve avoiding a greater harm. This is what Shiffrin calls a *pure benefit*. So, the idea is that it is impermissible to confer a pure benefit on someone who has not consented to it, while it is permissible to confer a benefit on someone to prevent a nontrivial harm to them. In the Rescue case there is hypothetical consent to the harm, whereas in the Gold manna case there is no such consent.[6]

The anti-natalist urges that procreation is analogous to the Gold manna case, not the Rescue case. Procreation imposes a nontrivial and unconsented harm on the individual who is created for the purposes of bestowing a *pure benefit*. Those who would procreate, then, do not have the hypothetical consent of the individuals they procreate. Why is this the case? If an individual does not exist, she cannot be harmed nor benefitted. Remember that language is misleading here because when procreation does not occur there is no 'individual' who does not exist. There is simply nothing. There is no person in a burning car, no people on the island, and no free-floating soul waiting to be created. Procreation always involves bestowing a pure benefit, something this argument says is impermissible.

These considerations lead to the following standardized version of the Hypothetical Consent Argument:

(13) Coming into existence can be an overall benefit or harm.
(14) All human lives contain some substantial harms.

[6] Benatar gestures at this argument, noting that prior consent cannot be obtained for procreation (2006, 99).

(15) Although it is often permissible to cause an non-consented harm to avert a greater harm, it is impermissible to cause an non-consented (nontrivial) harm to procure 'pure benefits' – goods that do not involve the removal or prevention of greater harm.
(16) It does not harm a merely possible individual not to bring her into existence.
(17) To procreate is to cause unconsented (nontrivial) harm in order to procure pure benefits.
Therefore,
(18) Procreation is impermissible. (Modified from DeGrazia, 2012, 151–152)

Objections and Replies

It is possible to agree that imposing a harm on someone in order to bestow a pure benefit is morally problematic but that it "may be permissible in some circumstances in view of other values" (DeGrazia, 2012, 153). If correct, this would show that (15) is false. A slightly different objection says that (17) is false because procreation does not typically impose significant harms "such as a major disability or a substantial socioeconomic disadvantage such as severe, entrenched poverty" (DeGrazia, 2012, 153). Instead, procreation is better understood as *exposing* their child to certain harms rather than *imposing* such harms on them. For example, DeGrazia contends that "[t]here is no harm closely associated with procreation in those circumstances, yet we know there will be some significant harm in that child's life. The difference between imposing harm and exposing to harm is significant" (2012, 153). The alleged significant difference is supposed to be illustrated by the following:

> [W]e faultlessly expose children to harm all the time, often as part of creating opportunities for great benefits. Parents who enroll their children in school expose them to any manner of possible harms: humiliation in class, despair at being dumped by friends, depression secondary to poor performance or social difficulties, emotional and physical bullying, minor physical harms during recess or gym class, sleep deprivation, illness

> contracted from other students, and so forth. Nor would it make much sense to say that the children consent to be so exposed, considering their immaturity during most of their primary and secondary school years. (DeGrazia, 2012, 153)

In sum:

> [B]ringing a child into the world where the child has good prospects in life and is not expected to have some major disability or hardship is not much like dropping gold cubes where they can land on someone. It is more like dropping gold cubes in the knowledge that passersby will pick them up and, simply by becoming wealthier, face an increased likelihood of being robbed and the emotional anguish of deciding which family members or charities to give to. (DeGrazia, 2012, 154)

If this is right, then there is reason to reject the Hypothetical Consent Argument for anti-natalism.

Notice that there are resources from the Risk-Based Argument that can be leveraged to respond to this objection to the Hypothetical Consent Argument. It is true that parents are not always *imposing* harms on their children but are instead imposing the risk of harms (i.e., exposure to them). Again, there is an empirical question about how frequently those harms occur. Benatar appeals to much data to show they are extremely likely to occur. Indeed, as stated above, the rate of cancer in Canadian men is over 45% and this is just one disease. Contra DeGrazia, the chances of getting struck by a golden brick is extremely high if the case is going to be close to procreation. Indeed, even if the risk of significant harms is much lower than Benatar supposes, a key difference between the cases DeGrazia mentions (as with the Prodigy Case above) is that the *child already exists*. Or a person who would benefit from the gold already exists. Once a child exists, it may well be reasonable to take some risks in order to gain benefits. I will not rehearse more of the details of the Risk-Based Argument from above; I simply note that shifting the Hypothetical Consent Argument from the imposition of significant harms to gain pure benefits to the imposition of the *risk* of significant harms to gain pure benefits does not help rescue the argument.[7] The Hypothetical Consent Argument can be successfully reframed as a Risk-Based Argument.

4.4 The Exploitation Argument

A different argument for anti-natalism appeals to the suffering involved in bringing about a fully formed human adult. I call this the Exploitation Argument and develop it based on ideas from Christopher Belshaw (2012).

The Argument

Consider that we have the intuition that we should end the lives of animals who are suffering even if there is some chance that they could be happy in the future (Belshaw, 2012, 120). Suppose that there are categorical desires, which involve reasons to ensure our future existence. Further suppose that there are also conditional desires, which, assuming a future, offer reason to think that one state of affairs will obtain over some other one (Belshaw, 2012, 121). Belshaw continues to suggest that while a baby is a human animal, it is necessarily *not* a person in a more robust sense. This is because babies are not psychologically knit together, nor do they have categorical or conditional desires (Belshaw, 2012, 124). Likewise, there is no continuity between a baby and the adult it becomes; it is implausible to think these are the same person. For a baby:

> [H]as no developed notion of itself, or of time, no desire to live on into the future, no ability to think about the pain and decide to endure it. Further, if we think seriously about a baby's life we'll probably agree it experiences pain in more than trivial amounts. Even perfectly healthy babies come into the world screaming, cry a lot, suffer colic and teething pains, keep people awake at night. None of us can remember anything about it. (Belshaw, 2012, 124)

---

7 Benatar (2012, 147) makes a similar point. See also Benatar (2012, 152, 157) and Smilansky (2012).

This sort of life is not, in itself, worth living. That having a baby exist is the only way for a fully formed human adult to be brought into existence does not in anyway compensate the baby for the harm it experiences (Belshaw, 2012, 124). The view that a baby gradually becomes a person does not help avoid the Exploitation Argument. At bottom there are still two distinct persons in the baby and the fully formed adult. A baby is necessarily exploited in order to achieve a fully formed adult. Belshaw suggests that because people tend to value their own lives, they will continue to have children. But even if procreation is good for adults because it eventually yields more fully formed humans, it does not follow that procreation is good for babies. Adults are exploiting babies when they procreate and indeed, only exist because of that exploitation. Not regretting the fact that one was a baby is consistent with the claim that it is better that babies are never born (Belshaw, 2012, 124).

It is true that one difference between a suffering animal and suffering baby is that most babies have the future of being a fully formed adult in front of them. However, no baby has any interest in becoming an adult. If the death of a baby is bad, then death is also bad for fetuses, embryos, and it is also bad even when conception is frustrated. However, "[o]nly a handful of those thinking, as altogether naturally most of us do think, that it is a terrible tragedy when a baby dies, want to embrace all this" (Belshaw, 2012, 125). Belshaw asks the reader to:

> Imagine we look at creatures on a distant planet. They live for ten years in agony, and thereafter sixty years in bliss. And there are no important psychological connections over this period. We probably think it better if these creatures never come to be, though of course after ten years there's no reason at all to kill them, or wish them dead. This is, I've argued, more or less the picture with human lives. (Belshaw, 2022, 126)

In sum, by procreating "we are inevitably free-riding on the several misfortunes of small, helpless and shortlived creatures" (Belshaw, 2012, 126).

Objections and Replies

A natural place to challenge this argument is on its controversial metaphysical claims. One may reasonably reject as absurd the claim that you or I were never babies (see Benatar, 2012, 163). Yet even if one makes this objection, it is possible to reinterpret this argument as providing additional data for the Quality of Life Argument. The harms associated with being a baby are not typically considered in discussions about the harms of existence, including in the work of Benatar. Furthermore, these are harms that *everyone* experiences, unlike many of the others often discussed. It is not guaranteed that any particular person will suffer from cancer, HIV, assault, poverty, but it is a guarantee that they will suffer from the harms associated with being a baby. Of course, they cannot be leveraged for the Risk-Based Argument since such harms are guaranteed. Having said that, there are probably ailments that are not guaranteed to occur but are exclusive to babies or much more likely to occur to them adults. These sorts of harms could indeed be leveraged in a Risk-Based Argument.

4.5 The No Victim Argument

Another argument for anti-natalism is offered by Gerald Harrison who argues that to coherently posit the existence of moral duties entails that there must be a possible victim (that can be hurt by the breaking of a duty). In light of this, he suggests that "we have a duty not to create the suffering contained in any prospective life, but we do not have a duty to create the pleasures contained in any prospective life" (2012, 94).

The Argument

It is intuitive to think that we have the following two duties: (1) There is a duty to prevent suffering; and (2) There is a duty to promote pleasure (Harrison, 2012, 96). Since there would be no victim if one failed to create happy people, this nicely explains why we do not have a duty to procreate even if we are sure our offspring will have very happy lives. However, this also explains why we have a duty not to create suffering people since if we do so there are clearly victims (i.e., the suffering people who were created). In other words, instead of an asymmetry between harm and pleasure, there is an asymmetry between moral duties.[8]

Since all lives contain suffering, there is a duty to never procreate. For in procreating, we always fail to do our duty to prevent suffering because there is an actual victim of suffering. That an individual has an on balance or overall happy life cannot outweigh the duty to not procreate because in failing to procreate there is no victim (Harrison, 2012, 97–99). According to Harrison, the duty not to procreate is therefore underpinned by two *prima facie* duties. These are prima facie duties along the lines of W. D. Ross such that they apply, all else being equal (Harrison, 2012, 95–96; See 203–204; Huemer, 2005, Ross, 1930, 19–20).

The first is the duty to prevent harm. Second, we have a "duty not to seriously affect someone else without [their] prior consent" (2012, 100; see 4.3). However, Harrison acknowledges that "[f]ulfilling this duty will mean that no more lives are created and this [...] is a bad state of affairs, even if it is not bad *for* anyone" (2012, 100). The reason we do not have a duty from ensuring that this state of affairs does not obtain is that doing so would involve bringing people into existence who will be harmed by their existence. On the other hand, that no more lives are created does not harm anyone. Harrison further notes that though his position entails the strange claim that a person can be happy for being brought into existence, even though they are harmed by it, there is nothing incoherent in it. For example, someone could place a large bet in our name without our consent. Doing so is wrong even if the bet is won and we ultimately benefit from it (Harrison, 2012, 100).

[8] Benatar suggests that the asymmetry here can be explicitly stated as: "The asymmetric counterpart to the requirement that there be a victim in the event that the purported duty is breached is the absence of a requirement that there be a beneficiary if the purported duty is fulfilled" (2012, 160).

Objections and Replies

One challenge for this argument, as observed by Benatar, is that it has no appeal to those who reject the notion of *prima facie* duties, including utilitarians, deontologists, and virtue theorists (2012, 161). Another is that even supposing there can be duties such that if an act is not performed there is a victim, it is easy to ask *why* this is the case. Benatar presses this point because he suggests that the asymmetry here might be better explained by his Basic Asymmetry (see 3.2). Benatar suggests that "Harrison's duty asymmetry tells us that there can be a duty only if there is a victim in the event that the purported duty is not performed. But why should that be the case? Why should it not be the case that there can be a duty only if there is a *beneficiary* in the event that the purported duty *is* performed?" (2012, 162).

4.6 The Argument from the Right to Physical Security

Blake Hereth and Anthony Ferrucci defend anti-natalism based on the Argument from the Right to Security (2022).[9] They begin by first defining the right to physical security (from here on 'RPS') as the claim that: All persons have a presumptive right that others avoid moral responsibility for unjust physical harms to them" (Hereth & Ferrucci, 2021, 17). Of course, RPS is a presumptive right because it can be forfeited, and it is a right that is sensitive to moral reasons. It is a right that moral agents "*avoid being morally responsible* for harms to which S neither consents nor is liable" (Hereth & Ferrucci, 2021, 17).[10] Their claim is that RPS entails anti-natalism.

---

[9] Coates makes a similar point in claim that parents are an accessory to the evil that their children experience (2014, 13–14).

[10] Hereth and Ferrucci claim their version of RPS is more attractive than others (see Hereth & Ferrucci, 2021, 17–19).

Their version of RPS does not say moral responsibility is strictly causal, which allows them to avoid problems in procreative ethics, including the non-identity problem:

> According to the non-identity problem, coming into existence cannot itself be a harm because it does not make anyone worse-off than they otherwise would be, since they would not exist otherwise and thus would lack a welfare altogether. Because we do not assume that coming into existence is *itself* a harm, our defence of anti-natalism avoids the non-identity problem. (Hereth & Ferrucci, 2021, 19)

This account of RPS also does not imply that procreators directly cause harm to their child. Parents do not cause their children to suffer cancer or HIV or other diseases.[11]

RPS implies anti-natalism because it is the case the one's children will experience significant or non-trivial harms. Hereth and Ferrucci suggest that it entails that most procreators are actually morally and criminally liable for harming those they create. Consider:

> *Newborn*: Blake has magical powers: With a snap of their finger, a new person appears out of nothing. But Blake's powers work only inside the holodeck, a place where the environment is always changing due to a randomised computer program. And Blake's creations cannot live outside the holodeck. Unfortunately, the safety protocols are always off, meaning that persons inside the holodeck are vulnerable to whatever environment is created. The odds of an environment resulting in non-trivial physical harms to a person are at least 80% for any given time someone is on the holodeck. As the holodeck comes to life, Blake snaps their fingers and creates a new person, Anthony, who at some point suffers a non-trivial physical harm. (Hereth & Ferrucci, 2021, 20)

Anthony's right to physical security is violated in *Newborn.* It is true that Blake is not responsible for the conditions of the holodeck, but Blake does bring Anthony into existence in those conditions. This is so even

[11] Hereth and Ferrucci argue that if RPS entails anti-natalism and the plausible right to physical security is true, then anti-natalism may turn out be enforceable (2022, 19). As stated in 1.4, I am not discussing issues of policy, so I will not focus on this aspect of their argument.

though Blake does not directly *cause* the harms that Anthony experiences. It is also not relevant that any particular harm is not guaranteed to happen to Anthonny (Hereth & Ferrucci, 2021, 20). Consider that harm is not guaranteed in Russian roulette, but that RPS is violated if someone is forced to play the game. It is not morally relevant whether the harms in question are delayed. Intentions do not also seem to matter because even if Blake does not intend to harm Anthony, Blake is still responsible for the harm. This means that "[e]ven if Blake was not aware of the risks to Anthony, and even if Blake believed Anthony's creation would result in pure bliss for Anthony, this makes no difference to the facts that there are serious risks [...]. And those facts generate an obligation for Blake not to create Anthony" (Hereth & Ferrucci, 2021, 20). Of course, *Newborn* is supposed to be analogous to procreation in the actual world.

This discussion leads Hereth and Ferrucci to offer a standardised version of the Argument from the Right to Security:

> (19) We should (other things being equal) avoid being responsible for non-trivial harms to persons to which they neither consent nor are liable.
> (20) If we create persons, they will suffer non-trivial harms to which they neither consent nor are liable.
> Therefore,
> (21) We should (other things being equal) avoid creating persons. (Hereth & Ferrucci, 2021, 21)

According to Hereth and Ferrucci, "[a] person is morally responsible for some harm if (a) the person freely performs an action that (b) either results in the harm or does not prevent it and (c) the harm was reasonably foreseeable (or should have been) by the person" (2022, 21). The significant harms that one's offspring will face are quite foreseeable (Hereth & Ferrucci, 2021, 22; see also 3.3). Admittedly, the majority of parents do not intend to cause their children harm by bringing them into existence. Hereth and Ferrucci write that "[p]arents typically act from beneficent or morally neutral motives when procreating. But their voluntary procreation, coupled with what they know (or should know) will almost certainly happen to their offspring, makes them morally responsible for the harms they foresee" (2022, 22). Thus, the only way for prospective procreators to avoid violating RPS to simply to refrain from procreating.[12] Finally, in referring to the Misanthropic Argument (see 3.4), they also note that one's children are also very likely to *cause* non-trivial harms and that their procreators are responsible for them.

Hereth and Ferrucci conclude by noting that they have shown there is a presumptive obligation to refrain from procreating but that not that procreation *all-things-considered* violates RPS. This means that "[p]ro-natalists can agree with this but insist that there are nonetheless justifications for transgressing RPS. That is, they will maintain that procreation *infringes*, but does not *violate*, the RPS of their procreated person or their victims" (Hereth & Ferrucci, 2021, 26). But a procreators' interests in becoming a parent cannot justify the significant harms associated with being brought into existence. Indeed, procreators seem willing to allow their children to experience great suffering in order to satisfy their interests in being a parent. It is true that since parents do not directly cause the harms, they are less wrong than they would be otherwise, but such interests hardly seem strong enough to justify violating RPS. To conclude, "the presumption established under RPS is sufficiently weighty to generate not only a presumptive obligation to refrain from procreation in almost all cases, but an all-things-considered obligation to do so" (Hereth & Ferrucci, 2021, 27).

An objector would be hard pressed to show that every person does not have their right to physical security violated at some point in their life. But even if they insisted that not every individual experiences it, there is a high probability that RPS will be violated during the course of someone's life. A high probability of violating RPS is enough justification to recommend refraining from procreation.

[12] They further appeal to Shiffrin to argue that parents are also *legally* liable for the harms their children experience.

4.7 The Argument from Negative Utilitarianism

Reference to negative utilitarianism is not always as prominent in discussions of anti-natalism as one might expect.[13] Of course, there are various approaches one could take with negative utilitarianism and so for the sake of simplicity, I am going to focus on a recent argument offered by Matti Häyry (2024). In its most basic form, classical negative utilitarianism is about minimising suffering as much as possible where this is equated something like pain (Häyry, 2024, 217; see Narveson, 1967). More recently, it is associated with one's interests being dissatisfied (Häyry, 2024, 217; see Häyry, 1994).

Häyry compares his theory to classical negative utilitarianism which is only concerned with harm experienced by sentient beings. This does not assign any moral significance to positive value. The version of negative utilitarianism that Häyry employs is likewise concerned with avoiding suffering but also acknowledges the importance of autonomy or ruling oneself, in addition to a deferral clause to be explained below (Häyry, 2024, 217). This deferral clause moves Häyry's theory beyond what classical negative utilitarianism since is adds a deontological side constraint on the ways in which disvalue can be minimized and value can be maximized (Häyry, 2024, 217).

Häyry notes that negative utilitarianism emphasizes the reduction of suffering or need frustration among all sentient beings. If suffering is the only relevant value, then the requirement to reduce it applies equally to all sentient beings (Häyry, 2024, 218). However, if the frustration of needs is in play, then sentience only partly matters because different beings have different needs, and this fact is morally relevant. For example, Häyry writes that "[m]ost human beings (and maybe other living beings) are autonomous agents. As such, they have a fundamental need to conduct their lives according to their own will or reason. The frustration of this need is, I posit, as bad as, and independent of, the need not to suffer" (Häyry, 2024, 218). Consider that this implies that

[13] Häyry and Sukenick note that Karim Akerma arguably pioneers this approach to anti-natalism at the turn of the twenty-first century (2024, 29; see also Akerma, 2000, 2017, 2021, 2024).

it is impermissible to end another person's life against their will in order to eliminate their suffering. If continuing to live is a choice, then living is a 'need based prima facie right' (Häyry, 2024, 218). If there are conflicts between fundamental needs then the case must be deferred in order to determine relevant duties using 'logic, common sense, and proportionality' (Häyry, 2024, 219).[14]

Negative utilitarianism implies that the self-extinction of the human species would be a good thing because ending sentience is the only known way to end suffering. If such extinction were voluntary, it is also consistent with Häyry's version of utilitarianism that focuses on needs (Häyry, 2024, 219). But that negative utilitarianism entails that extinction would be a good thing has been levelled as a *reductio* against the view (Häyry, 2024, 219; see Smart, 1958). Häyry claims that his version of utilitarianism avoids this challenge because it comes with the deontological side constraint of respecting autonomy. The autonomy constraint explains why voluntarily ceasing to procreate is good, but it cannot be made into a moral duty for everyone:

> According to sentiocentrism [classical negative utilitarianism], someone with the proper means could have a duty to end the life of the species. If a person could, by pushing a Big Red Button, make humankind instantly disappear, the person would have a moral obligation to do so. Within my need-based version, the situation is different. (Häyry, 2024, 220)

On Häyry's view, if humanity collectively accepts that continuing to procreate results in fundamental needs frustration, then there is no duty to end the species. This means that:

> The Big Red Button does not enter the picture at all because its potential pusher is not involved in the autonomous decisionmaking of a humankind that has chosen to continue its existence. If there are dissenters, they can be allowed to end their own lives but not the lives of those who want to survive. (Häyry, 2024, 220)

[14] It is not entirely clear what he means by these three things.

Anti-natalism is a way to end suffering and need frustration (Häyry, 2024, 221). Procreation just contributes more cases of need frustration. Häyry's point is just that his version of negative utilitarianism does entail that anti-natalism must be forced on others (Häyry, 2024, 222).

Accepting negative utilitarianism, including Häyry's modified version of it is hardly uncontroversial. I will not belabour the various objections to it, but simply note that this style of argument is available to those would like to merge negative utilitarianism and anti-natalism, while avoiding the implication of pro-mortalism or speciecide. As I will suggest below (4.8), this is one argument for anti-natalism that is inconsistent with Christianity and so I will not appeal to it moving forward.

4.8 The Prima Facie Consistency of the Arguments with Christianity

All of the arguments just presented are at least *prima facie* consistent with Christianity, with the exception of the one based on negative utilitarianism (4.7). Consider Risk-Based Arguments for Anti-Natalism (4.2). Indeed, Christians should (all else being equal) want to avoid imposing the risk of harms on non-consenting innocent persons. This is importantly different from claiming that a Christian should herself avoid taking risks; the Christian ethic may indeed require a person to take on significant risk of harm. However, that is distinct from imposing the risk of harm on someone else. In Part III, I suggest that risk-based arguments offer some of the most promising ways of establishing that Christians should be anti-natalists, particularly those who are not universalists.

As with Risk-Based Arguments, the Hypothetical Consent Argument (4.3) is at least *prima facie* consistent with Christianity. Bestowing a pure benefit on someone that causes them significant harm or imposes the risk of significant harm is impermissible if the Christian ethic is true. This is so even if it could be argued that autonomy or consent are not as important on the Christian tradition as they are in, say, a secular humanist one. I leave it to others to show how these are incompatible. The same can be said for exploiting other people (4.4) (even to get a desired end), intentionally creating victims (4.5), and violating a person's right to physical security (4.6). Now, it can be objected that there are reasons to ignore consent, create a victim, or violate a right to security. But those would constitute objections to the arguments themselves. My point is just that *if* the arguments in question are sound, then they are *prima facie* consistent with Christianity.

Finally, negative utilitarianism (4.7), and even Häyry's modified version that contains a deontological constraint is incompatible with Christianity. Christianity does not only value the absence of suffering and the respecting of autonomy. For example, Christianity values a number of character traits that are directly about relieving suffering or exercising autonomy. Perhaps there are ways of showing that there is consistency between Christianity and negative utilitarianism, but I will not attempt to do so here. As such this is where I leave arguments for anti-natalism that appeal to negative utilitarianism.

4.9 Conclusion

The purpose of this chapter was to explain important arguments for anti-natalism that are primarily located outside the work of David Benatar. A common thread that emerged across many of these arguments involve risk, and the lack of consent to harm or potential harm. The Risk Argument explicitly involves weighing the likelihood of risk, while the Hypothetical Consent Argument can be formulated as one about consenting to the risk of harm instead of guaranteed harm. Part of the Exploitation Argument can be understood as that a baby cannot consent to the harm they experience (and not reap the future benefits associated with their current distress). There is also a lack of consent to becoming a victim or having one's right to security violated. These arguments also rely on the familiar claim in Benatar's work that no one is harmed by not being brought into existence, and so it is better not to procreate. With the background of the last three chapters now in view, I am finally ready to describe what I take to be the best arguments for Christian anti-natalism.

Part III

Christian Anti-Natalism

5

Anti-Natalism for Christian Exclusivists

5.1 Introduction

Everything up to this point in the book has been stage-setting for this chapter and the next one. I am now in a position to offer what I think are the most promising set of arguments for the claim that Christians should be anti-natalists. In this chapter, I develop three different risk-based arguments for anti-natalists that are intended to apply to Christians who are exclusivists (5.2). This argument says for Christians who believe (i) in a literal hell and (ii) that each person born could end up in hell, that procreation is impermissible. The risk of going to hell should not be imposed on a non-consenting person. Of course, just what constitutes hell and the likelihood one ends up there play a role in the force of this argument. I conclude that on traditional understandings of hell even a very slight risk of ending up there makes procreation impermissible. I further argue that the scope of this argument applies well beyond Christian exclusivism; anyone who believes there is a chance their offspring could end up in hell ought to refrain from procreating. After this I briefly examine some of the more obvious objections to the argument including those that say hell is either not that bad or is imposed justly (5.3). To conclude, I explore the extent to which some of the other arguments for

K. Lougheed, *Christian Anti-Natalism*, https://doi.org/10.1007/978-3-032-03902-6_5

anti-natalism can be modified in order to develop unique arguments for anti-natalism that apply to Chrisitan exclusivism (5.4).

5.2 The Risk-Based Argument for Christian Exclusivists

In this section, I develop a series of risk-based arguments for anti-natalism which applies to those Christians who are exclusivists and believe in the traditional doctrine of hell. I begin by outlining a general risk-based argument before showing how it applies to Christian exclusivists who assign different probabilities to the chances of ending up in hell.

The General Argument from Non-Trivial Risk

In 4.2, I suggested a plausible Risk-Based argument for anti-natalism could be formulated by endorsing the following morally intuitive principle:

> *Non-Trivial Risk Principle:* it is impermissible to non-consensually impose a risk of catastrophic harm on others when there is a non-trivial chance of occurrence, and there is insufficient justification for the imposition of that risk of that harm (which includes justification based on a non-trivial chance of advancing a third party's essential-interests or significant non-essential interests).

While this principle enjoys high intuitive plausibility, it is immediately vague with respect to what constitutes a 'non-trivial' chance. I make no claims to offer a precise account of 'non-trivial' chance, however I note that commonsense judgments suggest that the odds of experiencing a harm do not have to be very high in order to generate the relevant intuition. Consider that while driving drunk only very slightly increases the chances of an accident, especially one a person is only marginally over the legal limit, most of us agree that it is still wrong to drive drunk (Himma, 2010, 190–191). The same is true of non-consenting second hand smoke even though occasionally inhaling second-hand smoke does not *significantly* increase a person's risk of cancer. People tend to have an extremely low risk tolerance for imposing any type of non-consenting harms.

The Non-Trivial Risk Principle lends support to the following general argument for anti-natalism:

> (1) It is impermissible to non-consensually impose a risk of catastrophic harm on others when there is a non-trivial risk of occurrence, and there is insufficient justification for the imposition of that risk of that harm. (modified from Benatar, 2022, 149)
> (2) Bringing a child into existence involves non-consensually imposing a non-trivial risk of catastrophic harm on that child, with insufficient justification for the imposition of that risk of that harm. (modified from Benatar, 2022, 149)
> Therefore,
> (3) It is impermissible to bring children into existence. (Magnusson, 2022, 106; *numbering mine*)

Benatar's formulation says that the risk of catastrophic harm must be 'highly probable' whereas I suggested that if the harm is significant, even a much lower probability of it occurring suggests imposing the risk of it is impermissible. I will say more about how different arguments can be constructed based on divergent risk tolerances below (see 5.2.3; 5.2.4).

The Argument from Non-Trivial Risk and Chrisitan Exclusivism

Before showing how this argument applies to Christian exclusivists, a number of definitions are in order.[1] Exclusivism is the view that in order to gain salvation a person must explicitly accept Christianity. There is much recent debate over the nature of faith (see Lougheed & Hendricks, 2021). Does genuine faith require assenting to a set of propositions? What about having certain dispositions or attitudes? Likewise, the notion of a 'nominal' Christian suggests that some (perhaps many) people who

[1] For more see Basinger (2020).

claim to be Christians are not in fact saved. I do not need to wade into all of these issues. The basic claim of the Christian exclusivist is that one must be a Christian (whatever one means by this) in order to be saved. Exclusivists need not take any particular view about the doctrine of hell. They could simply present a position on what is required for salvation and remain silent on the rest. However, when I write of exclusivism in what follows I attach to it the traditional doctrine of hell.[2] This doctrine says that there is a literal (spiritual) place where persons who are not saved experience the worst possible sustained conscious torment for all of eternity.[3] This also tends to be combined with the view that hell is in fact populated. In short, by exclusivism I mean the view that only Christians are saved, not everyone is a Christian, and hell is a place of torment where all non-Christians spend eternity.

Exclusivism can be contrasted with the two competing views of inclusivism and universalism. The former of these is especially slippery but I will take it to mean that view that not every person who receives salvation must explicitly accept Christianity. Perhaps one needs to live a sufficiently good life, or sincerely adhere to another religious faith, etc., in order to eventually receive salvation through the Christian atonement. Or perhaps one does not need to accept Christianity but receives a second chance in the afterlife to accept it (something I stipulate the exclusivist denies). Or maybe the relevant category is about counterfactuals such that it needs to be the case that in a certain set of circumstances the person would have been a Christian, even if those circumstances never obtain. Finally, universalism is the view that every person (eventually) receives salvation. Of course, universalists may disagree about how the mechanics of this occur, but the point is that in the final analysis everyone will gain salvation. With these definitions in mind, I can now offer the Argument from Non-Trivial Risk and Christian Exclusivism:

---

[2] In writing on the same issue, Kenneth Himma pulls them apart (2010, 2016).

[3] Indeed, Himma develops a risk-based argument along these lines, though one that is also reliant on the Basic Asymmetry (2010). However, his goal is to show that either exclusivism or the doctrine of traditional hell must be false, since he assumes that procreation must be permissible.

(4) It is impermissible to non-consensually impose a risk of *hell* on others when there is a non-trivial risk one will end up in hell, and there is insufficient justification for the imposition of that risk of that harm.
(5) Bringing a child into existence involves non-consensually imposing a non-trivial risk of hell on that child, with insufficient justification for the imposition of that risk of that harm.
Therefore,
(6) It is impermissible for the Christian exclusivist to bring children into existence.[4]

Premise (4) is, I think, firmly established in light of the discussion in 4.2. Hell is a catastrophic or significant harm since it is *by definition* the worst thing that could happen to a person, especially on the assumption that it is *eternal*. If it is not considered a significant harm, then significant harms simply do not exist. In light of this fact, it is intuitively obvious that it is impermissible to impose the risk of the *worst harm possible* on a person with insufficient justification. There must be very compelling reasons for imposing such a harm.

It is worth pausing to spend some time examining the notion of 'non-trivial' risk in the case of hell. Suppose the risk of hell was 0.0000001 or that hell is only populated with a handful of people. Universalism is still technically false on this view, but it might be thought that if the odds of ending up there are miniscule, then procreation is permissible (i.e., imposing a miniscule risk on a non-consenting person is allowed). Perhaps this is why Benatar's formulation of a risk-based argument says 'high probability of occurrence' (see 4.2.2). But I am concerned here with the traditional doctrine of hell and so this changes the terms of the argument. The traditional view does not say that hardly anyone is assigned to hell. Likewise, even if the language of 'non-trivial' is supposed to eliminate cases where the risk of harm is infinitesimally small, consider that

4 In one of the few places to explicitly engage Christianity and anti-natalism, Ben R. Crenshaw examines what could be also be construed as a risk-based argument which says "it is categorically wrong for parents to impose the harm of existence upon their children (something parents do when they procreate), since there is no guarantee that the child will be redeemed by Christ, which would make their life worth living" (2018, 5). He appears to ultimately reject this argument because the life, work, and suffering of Jesus justify the suffering we experience in life (Crenshaw, 2018, 152–153).

Augustine held that only 10% of churchgoers in his day were actually saved. On whatever criteria Augustine used for salvation, is there any reason to think his estimate would be much higher at different points in human history?

As I will discuss below, it is doubtful that even if the probability of ending up in hell is extremely low that this argument can be ignored. But there is also empirical data that is relevant. One question about probability concerns how likely it is that hell exists. Yet another question is how likely it is one's children will end up in hell, on the assumption there is one. I will not focus on this topic as I worry that even slightly contested empirical data detracts from my main point. However, in discussing the risk of hell and procreation, Bruce Blackshaw observes that "according to Bengtson [et al.], a majority of today's younger generations—around 60%—retain the beliefs of their parents (2013, 186), and this proportion has remained stable for decades" (forthcoming). This means that around 40% do *not* retain Christianity. Indeed, a Christian of any experience will have witnessed children who grow up with Christian parents but do not continue as Christians into adulthood. I mention this only because it is implausible to respond that even if there is a hell that prospective parents could guarantee that their own children would avoid it.

As mentioned, also is unclear that the argument fails if the risk of ending up in hell is extremely low. The analogy between procreation and Russian roulette is instructive. Kenneth Himma suggests that when the harms in question are extremely great, a normal cost-benefit analysis cannot be conducted. He explains that:

> The problem here is that consignment to eternal torment is as catastrophic an outcome as is possible for any person from the standpoint of prudential rationality. If I were temporarily given rationality while in the womb and were informed of the relevant epistemic probabilities of the potential risks and rewards of being born, I would choose not to be born even if the chances of eternal bliss are 5 out of 6 and eternal torment just 1 out of 6. While it might be true that eternal bliss is as positive an outcome as is possible from the standpoint of prudential rationality, I would be no more willing to play Heavenly Roulette with a 5 in 6 chance of winning than I would be to play Russian Roulette with those odds—no matter how desirable the prize is. (Himma, 2016, 104)

Even if the odds of ending up in hell are significantly lower than this, I suspect that most people would not take the risk themselves. Perhaps there are odds such that the risk of hell is so minuscule that it would be irrational not to risk it when the other outcome is eternal bliss. Perhaps there is a literal hell but the odds of any individual ending up there are so small that no individual in fact ends up in it. Again, this is not the traditional view of hell, and given the type of risks we are normally unwilling to impose on others (e.g., a very low risk of lung cancer or a car accident), it remains impermissible to procreate if exclusivism is true.[5]

With respect to premise (5), the reasoning offered in 4.2.2 and 4.2.4 suggesting that the type of risk humans ordinarily tolerate show there could be no justifying reasons for imposing the risk of hell on another non-consenting person. The interests of the child can never be promoted by being brought into existence because non-existent beings do not have interests. The interests of the parents (or some other third party), essential or non-essential, also cannot justify imposing the risk of hell on someone else. Recall that if it is wrong to impose the risk of cancer by smoking in the presence of someone who is non-consenting, it is wrong to impose the risk of the worst possible harm on a non-consenting person. This is so even if the risk of cancer comes with gaining some essential or non-essential interest. Consider that you could not impose the risk of cancer on another couple because you want a child. Since hell is far worse than cancer, it is difficult to see how the force of the reasoning here is not even stronger.

(6) follows from the truth of (4) and (5). Christian exclusivists act immorally when they procreate. This is a particularly interesting result because I suspect that Christian exclusivism is correlated with pro-natalism. Of course, this is a sociological claim I cannot defend here, but it does point to the possibility of interesting future research in the sociology of religion.

[5] I make no attempts here formalise this precisely using decision theory. Whatever formal theory one wants to use for this analysis, it must accord with our commonsense judgements about risk imposition.

The Argument from Unknown Risk and Christian Exclusivism

Suppose that someone insists the risk of ending up in hell is unknown. I submit that there is a parallel argument that is perhaps even stronger than the Argument from Non-Trivial Risk. Consider a different risk principle that appeals to an unknown risk of some harm obtaining:

> *Unknown Risk Principle:* it is impermissible to non-consensually impose a risk of catastrophic harm on others when there is an *unknown* chance of occurrence, and there is insufficient justification for the imposition of that risk of that harm (which includes justification based on a non-trivial chance of advancing a third party's essential-interests or significant non-essential interests).

The 'unknown' here needs clarification. It is not intended to imply to anything that is logically possible such that there is an unknown chance I could end up spending eternity as a ghost inhabiting a moon made of cheese after my biological death. Instead, I mean that it is known that the harm occurs, but the likelihood that it occurs is unknown. If it is impossible to know whether there is a say, 1% or 99% chance, of imposing the risk of a catastrophic or significant harm, it is impermissible to impose it. Suppose you knew there was a chance that your child could end up in perpetual sexual slavery, but you did not know the precise odds. You only know that it is neither 0% nor 100%. The odds could therefore be quite low or also quite high. I take it as intuitively obvious you should not risk having a child in such circumstances. Since hell on the traditional view is worse than sexual slavery, this principle applies to the case in question too.

Someone might respond that upon reflection, prospective parents know perfectly well this fate might befall their children but remain undeterred presumably because they deem the risk of such a terrible fate to be sufficiently low. But that people impose such risks on other is not evidence that they are justified in imposing them. Even if I am wrong, the doctrine of traditional hell does not place the odds of ending up in hell as miniscule, even if the precise odds are unknown.

In reflecting on the fact that it is impossible to know whether someone is saved, Kenneth Himma writes that:

> [I]t is simply not possible to be epistemically justified with respect to any theory of what constitutes authentic saving Christian faith. There is a striking range of disagreement on this issue, which should be enough to show that we do not have enough information on the question that would clearly show which view is correct, and hence that no one is epistemically justified in whatever view they take on the issue. (Himma, 2010 197)

Thus, an even stronger argument for anti-natalism for Christian exclusivists is:

> (7) It is impermissible to non-consensually impose a risk of *hell* on others when there is an *unknown* risk one will end up in hell, and there is insufficient justification for the imposition of that risk of that harm.
> (8) Bringing a child into existence involves non-consensually imposing an *unknown* risk of hell on that child, with insufficient justification for the imposition of that risk of that harm.
> Therefore,
> (6) It is impermissible for the Christian exclusivist to bring children into existence.

If you knew that second-hand smoke increases the chance of getting cancer or that driving while drunk increases the chance of getting into an accident but were unsure how much it did so, it would remain impermissible to impose the unknown risk on a non-consenting person. All of this is likely consistent with what many exclusivists believe about salvation. They will probably admit that the risk of ending up in hell is ultimately *unknown.* The traditional doctrine of hell stipulates that there is a populated literal hell. But this says nothing very specific about the chances one's offspring will end up in hell. Part of the problem here is an epistemic one; it is simply impossible to know whether any particular individual is saved prior to their death. There is no way to check or confirm where a person ends up when they die.

The Master Argument from Risk and Expanding Its Scope

Thus far I have not offered a standardized version of this style of argument that has the risk of hell as 'significant' since it is less controversial than the two iterations that appeal to 'non-trivial' or 'unknown' chances of harm. If they succeed, then an argument appealing to a significant chance of harm will also succeed. These ideas can be synthesized into the following Master Argument from Risk:

> (9) It is impermissible to non-consensually impose a risk of *hell* on others when there is a *significant, or non-trivial, or unknown risk* one will end up in hell, and there is insufficient justification for the imposition of that risk of that harm.
> (10) Bringing a child into existence involves non-consensually imposing a *significant, or non-trivial, or unknown* risk of hell on that child, with insufficient justification for the imposition of that risk of that harm.
> Therefore,
> (6) It is impermissible for the Christian exclusivist to bring children into existence.

In sum, whether the risk of ending up in hell is significant, non-trivial, or unknown, it is impermissible for the Christian exclusivist to procreate. Though I have framed this argument with respect to Christian exclusivism here, this argument actually extends well beyond exclusivists. For example, many other Christians are not epistemically certain that exclusivism is false and so would have to admit that there is a chance their offspring end up in hell.[6]

Agnostics, typically understood, are unsure about whether God exists.[7] But someone could also be agnostic about whether Christianity in general or Christian exclusivism in particular is true. I suspect that at least

---

[6] See Blackshaw forthcoming for a related about credences in the context of a discussion about hell and the permissibility of procreation. Nathan Stagg has recently argued that if Rawls Maximin decision principle is used, then from the Original Position procreation is impermissible because the risk of hell is unknown or insecure (forthcoming.

[7] They could think the evidence for and against God's existence is on balance even. Or they could think the current evidence is inclusive such that they should suspend judgment.

some committed atheists who are very confident that God does not exist will also admit Christian exclusivism is a possibility, even if only a small one.[8] The group of people who believe hell is *epistemically possible* is presumably far larger than just Christian exclusivists. If a person believes hell is epistemically possible, then they ought to refrain from procreating because it is immoral to impose the risk of the worst possible fate on a non-consenting person. This means that (6) should be widened to:

> (6*) It is impermissible for anyone who believes the risk of hell is *significant, or non-trivial, or unknown* to bring children into existence.

Up until now, I have intentionally refrained from stating this argument in terms of what individuals *believe* because I fear it risks creating confusion and distracting from the main claims of these arguments. However, the argument is really supposed to apply to individuals with certain beliefs. Christian exclusivism as I have defined it, is a position about what a person believes about death and the afterlife, etc. It applies to those individuals who hold the set of beliefs identified with exclusivism. I claim that it applies to a further set of people who hold a different set of beliefs about death and the afterlife. Namely, they assign a lower credence to the existence of a populated literal hell.[9] Recall that one way to understand my aim in this book is to explore the extent to which there can be reflective equilibrium in holding both Christianity and anti-natalism are true. It is fundamentally about whether certain beliefs (i.e., those associated with Christianity) require anti-natalism. I have now argued that not only do these arguments apply to those Christians who are exclusivists, but it also applies to anyone who believes their offspring *could* end up in hell.

---

[8] Blackshaw forthcoming explicitly says this argument will apply to atheists who given a slight credence to the possibility of hell.

[9] It is not necessary for me to choose between the tripartite understanding of belief or credence understanding of beliefs (see Schwitzgebel, 2024). I also do not need to specify the strength of credence required for the label 'belief'. I am optimistic that this argument can be reframed with whatever model one wants to employ.

5.3 Objections and Replies

In this section I briefly consider a number of objections to these arguments. These include the claim that hell is not a significant harm, children cannot exist in a Christian exclusivist world without the risk of that harm, and that unlike many other harms, hell is imposed justly.

Hell is not a catastrophic or significant harm

One objection to the Risk Based Argument for Christian Exclusivists is to simply deny that hell is a catastrophic or significant harm. Some models of hell suggest that hell is just the *absence* of God's presence or of the beatific vision. The harm involved in hell is therefore this absence and any of the harms that come with it (for more see Buenting, 2010; Lewis, 1945; Walls, 1992). Hell is thus not conscious eternal torment and imposing the risk of it on one's offspring is permissible.

I leave it as an open question whether it would be permissible to impose this kind of risk on a non-consenting individual. However, notice that this is *not* the traditional view of hell that the argument employs. I am *stipulating* that hell is a place of eternal conscious torment and that it is populated. There are many Christians today who hold this view, or something close to it. The argument simply does not apply to those who do not believe in this version of hell.[10]

The Child Would Not Exist Without the Imposition of Risk

Another objection is to say that for the Chrisitan exclusivist, no person can exist without the risk of hell. Perhaps some different person could exist in a different possible world, but no individual in our world could exist without the risk inherent in hell. However, it is difficult to see why this justifies the risk, instead of just reinforcing the strength of the

[10] Though notice that if a less bad version of hell constitutes a harm, many of the arguments for anti-natalism apply to those who believe its possible.

case for anti-natalism. Suppose there was a one in six chance that your offspring would be born into permanent sexual slavery. You cannot have offspring without imposing this risk on them. It is true that, perhaps, were this risk not to exist, the person created would be a different one (i.e., created in a different possible world). Still, it is difficult to understand why the fact that no person would exist without the imposition of this risk somehow justifies bringing that person into existence. One need not accept the Basic Asymmetry (3.2) in order to see that a non-existent being cannot be harmed (i.e., one can disagree with Benatar about whether the absence of good is not bad). No one is harmed by not being brought into existence.

Hell Is Imposed Justly

A different objection says that there is a disanalogy between imposing the risk of a disease or injury on someone with the risk of hell since the latter is imposed justly (Bawulski, 2013). Shawn Bawulski argues that "No respectable account of [traditional hell] posits reprobation as a direct result of being born. Being born does not decide the probability or actuality of one's eternal destiny: some other factor does" (Bawulski, 2013, 8). Sin, not existence, is the reason why a person ends up in hell. This explains why "[u]sually hell is depicted as the result of the individual moral agent's sin and lack of repentance, never something that "happens to them" in a way an illness does" (Bawulski, 2013, 338). Though a person is not typically morally responsible for getting a harmful disease, they are responsible if they end up in hell.

However, this objection assumes that *causation* is playing an important role in being born and ending up in hell (see Himma, 2016, 100). The Risk Principle I appeal to does not use the word *cause.* It is still the case that being brought into existence puts one into a position where someone could end up in hell, even if they are culpable for that once they are born. The risk of being culpable could altogether be avoided (Himma, 2016, 101). Consider whether it is of any comfort to imagine one's child in hell but with the knowledge that the child is responsible for being there. Suppose that you knew, prior to conceiving, there was a non-trivial chance (i.e., it happens to some people's children) your child would commit crimes and end up serving a jail sentence of infinite length of conscious torture. Even supposing (perhaps implausibly)[11] that the punishment fits the crime, one should still not subject their offspring to the risk of such punishment.

5.4 Additional Arguments

I believe the above Master Argument from Risk provides a compelling case for Christian exclusivists to be anti-natalists, and indeed also those people who believe hell is a possibility. There are no reasons that could serve to justify imposing the risk of hell on one's offspring, at least when the traditional doctrine of hell is in view. Though these sorts of considerations are underexplored in the philosophical literature on anti-natalism, I believe they constitute some of the strongest reasons for the view. Having said that, in this section I turn to exploring whether some of the other arguments for anti-natalism discussed in Chapters 3 and 4 might also be reformulated to apply to Christian exclusivists.

Benatar's Other Arguments

I now briefly explore whether any of Benatar's arguments apart from his gesture at a risk-based one, could be leveraged to provide unique arguments that apply to Christian exclusivism. The Asymmetry Argument says that because the absence of harm is good while the absence of pleasure is not bad, procreation is always impermissible (3.2). A person always experiences some harm at various points in their life but is alternatively not harmed if they are not brought into existence. It is true that bringing hell into the picture adds a significant harm – the worst one possible – but it does not yield the material for a unique argument based on the Basic Asymmetry. Experiencing any harm at all is enough to justify anti-natalism according to this argument. The existence of hell

[11] See Lougheed (2024).

does not add any new information to help determine whether the Basic Asymmetry is true. Also, the harm of hell is not one that is guaranteed to occur and so it is not necessarily a harm that would be relevant to the Basic Asymmetry.

The Quality of Life Argument (3.3) does not consider those assessing the quality of their lives from the perspective of hell. Again, not every person ends up in hell, and Benatar is considering the quality of people's lives apart from any considerations about the afterlife. Finally, the Misanthropic Argument (3.4) does not point to any unique route for establishing anti-natalism for the Christian exclusivist. In sum, though appealing to the concept of hell does not show any of Benatar's arguments are unsuccessful, they do not point to any unique ways the arguments could be leveraged, unlike Benatar's gesture at a risk-based argument which I appeal to in the above (5.2).

The Hypothetical Consent Argument and Christian Exclusivism

Recall from the previous chapter that the Hypothetical Consent Argument (4.3) says that it is impermissible to cause a non-consented to and non-trivial harm in order to procure 'pure benefits'. Since an individual would not be harmed by not being brought into existence, the only reasons for bringing someone into existence must be described as 'pure benefits'. So, it is impermissible to procreate. The argument hinges on the sort of harms that humans do in fact experience. Here again the risk of experiencing hell can be added to the discussion but the argument is not necessarily strengthened by it. If every person were guaranteed to end up in hell, then it would certainly strengthen the argument. It would be forcing an unconsenting harm (i.e., the worst harm possible) on someone in order to gain the pure benefit of earthly existence. Perhaps imposing the risk of a non-consenting harm is a type of harm in itself which is similarly unjustified. However, if this is right, then the considerations that apply to the Risk-Based Arguments also apply here (5.2) and they need not be repeated.

The Argument from the Right to Physical Security and Christian Exclusivism

Similar considerations that were brought to bear on the Hypothetical Consent Argument also apply to the Argument from the Right to Physical Security regarding hell. The slight difference is that the latter emphasizes the responsibility of the parents for the harms in question. Procreating means inevitably being responsible for violating a person's right to physical security. Again, on the Christian exclusivist picture I am dealing with, not everyone ends up in hell. So, the responsibility for ending up in hell does not inevitably fall on all parents. However, on this view all parents risk being responsible for inflicting the worse possible outcome on their child. This is significant. Although an individual ends up in hell through their own decisions (see 5.3.4), their parents put them into the situation where they could up in hell in the first place. This means that parents bear at least partial responsibility if their child ends up in hell. To see this, consider a modified version of the Newborn (4.6) case:

> *Newborn**: Blake has magical powers: With a snap of their finger, a new person appears out of nothing. But Blake's powers work only inside the holodeck, a place where the version of Christianity that is true is always changing due to a randomised computer program. And Blake's creations cannot live outside the holodeck. Unfortunately, the safety protocols are always off, meaning that persons inside the holodeck are vulnerable to whatever version of Christianity is created. The odds of an environment resulting in Christian exclusivism (and the traditional doctrine of hell) are at least 80% for any given time someone is on the holodeck. As the holodeck comes to life, Blake snaps their fingers and creates a new person, Anthony, who is in a Christian Exclusivist world and so exposed to the risk of ending up consciously tormented in hell eternally. (Modified from Hereth & Ferrucci, 2021, 20)

Since Blake is responsible for Anthony being brought into a Christian Exclusivist world, then Blake is also at least partially responsible if Anthony ends up in hell. Notice that this argument can easily be modified to show that it applies to individuals who are not exclusivists.

Suppose the chance that the environment someone is born into will be a Christian exclusivist one is 50% which might be what other Christians who are not confident about eschatology or perhaps agnostics assign to it. Maybe a confident universalist would assign a 15% chance and a confident atheist only a 1% chance. The point is that in any of these cases, the person who chooses to press the button knowingly risks creating a person in a Christian exclusivist world. So, they risk violating the person's right to physical security and if such a violation occurs, they are indeed at least partially responsible. Of course, this is closely analogous to prospective parents.

A New Exploitation Argument?

The Exploitation Argument (4.4) appeals to the fact that babies have to be exploited in order for there to be fully formed adults. I noted that this is metaphysically controversial inasmuch as it implies that no adult was ever a baby. A different type of exploitation argument emerges in the context of exclusivism and hell. Consider that you might be exploiting a person who ends up in hell in order to have a child. Suppose you are deciding whether to have children and you are not as averse to risk as the arguments above suggest you ought to be regarding hell. You know that some people will end up in hell but are confident that your children will avoid this future. So, you have children, and you turn out to be right. None of them end up in hell. There is a sense in which you have exploited the people who end up in hell in order to have children. You knew that, given exclusivism, *some people* would end up in hell. This might be thought of as a kind of *systemic evil* because it is a bad-making feature of the world. You participate in this system by procreating even though your children do not experience the greatest harms that occur within the system. I do not endorse this argument and will not say more about it. But it strikes me that there are potentially analogous arguments for anti-natalism to be made to those about systemic injustices.

5.5 Conclusion

In this chapter I developed an argument for anti-natalism that applies to Christians who are exclusivists. In this context, by exclusivism I mean the view that one must be saved in order to avoid hell, where hell is populated and consists of eternal conscious torment. The main thrust of the argument is that it is immoral to impose a non-trivial risk of the worst possible harm on a non-consenting adult. What constitutes 'non-trivial' is admittedly vague. However, I showed that our commonsense risk aversion is actually quite low. We typically think it is impermissible to impose even slightly higher risks of cancer or car accidents on non-consenting people. So, it is impermissible to impose the risk of hell, the worst possible harm, on non-consenting people, even if one believes that the risk of doing so is relatively low. Further still, even if the risk of the harm is unknown, it should not be imposed on a non-consenting person. This argument most obviously applies to Christian exclusivists as I have defined them, since they posit a literal, populated, hell of conscious eternal torment. However, it also applies to anyone who believes that their offspring could possibly end up in hell. This likely includes most other Christians who are not exclusivists, but not epistemically certain in their rejection of exclusivism. Interestingly, the argument also likely applies to at least a subset of atheists and agnostics. In sum, it is impermissible for Chrisitan exclusivists to procreate since they should not impose the risk of the worst possible harm on a non-consenting person.

6

Anti-Natalism for Christian Universalists

6.1 Introduction

In this chapter, I examine a series of arguments for anti-natalism on the assumption that Christian universalism is true. I argue that the Asymmetry Argument, the Misanthropic Argument, the Exploitation Argument, or Argument from the Right to Physical Security are *not* defeated by universalism (6.2). Christians still need to grapple with these arguments in their own right, even if universalism is true.

I then explore two pro-natalist arguments which emerge from the assumption of Christian universalism since everyone's offspring will end up in heaven. The first of these is a reformulated quality of life argument that takes into account the fact that everyone ends up in heaven (6.3). I expand the scope of this argument by showing that the great good of heaven does not automatically make every life worth starting. The second focuses on the claim that every human created is intrinsically and finally valuable (6.4). I suggest that one potential problem for this argument is the absurd implication that humans should procreate as much as possible.

I conclude by offering what I take to be the strongest argument for anti-natalism on the assumption that Christian universalism is true. This

K. Lougheed, *Christian Anti-Natalism*, https://doi.org/10.1007/978-3-032-03902-6_6

is a risk-based argument which focuses on the idea that if there is a chance that one's offspring will be deprived of heaven, where the alternative afterlife outcomes are harmful, then procreation involves forcing the risk of a significant harm on a non-consenting person (6.5).

The Anti-Natalist Arguments and Christian Universalism

As stated in 5.2.2, universalism is the view that every individual (eventually) receives salvation. The mechanism about how this occurs is not important. Likewise, the content of what constitutes a salvific afterlife are not important. It could be eternal presence in the beatific vision or some other oneness in the presence of God. The point is that it is the best possible thing that could happen to a person. Of course, this is starkly different from exclusivism where ending up in hell is a live option, if not probable outcome. In this section, I examine some of the arguments for anti-natalism in order to discover whether they remain viable on the assumption that universalism is true. My aim is not to evaluate these arguments in their own right, but to discover whether universalism provides an obvious defeater for them.

The Asymmetry Argument and Christian Universalism

The Asymmetry Argument (3.2) does not appear to be defeated by the truth of Chrisitan universalism. The harms that Benatar outlines as part of any of human life do not disappear just because universalism is true. Indeed, all of the harms that Benatar is concerned with are those experienced in one's earthly life. The key to the argument, which is the Basic Asymmetry, is not undermined by the existence of a guaranteed good afterlife.

It is tempting to respond that the great good of a positive afterlife is so great that it can *outweigh* all of the harms people experience in their earthly lives. However, the Asymmetry Argument does rely on—indeed, does not allow for—the weighing of harms against pleasures (or benefits) in order to determine whether procreation is permissible.[1] Remember that if the Asymmetry Argument is sound, then if a person experiences *any* harm, it is better that they do not exist because they would not be harmed by not experiencing any of the benefits of existence. This includes failing to experience an exceedingly good afterlife. Non-existent people do not exist and so cannot be harmed by missing out on benefits, including blissful afterlives.

I have neither endorsed nor rejected the Asymmetry Argument. My point here is adding universalism as a point of consideration does not affect whether the argument is successful. Indeed, no view of eschatology will change the merits of the Asymmetry Argument since any individual will experience harm during their earthly life. Thus, Christians must address the argument on its own terms.

The Misanthropic Argument and Christian Universalism

The Misanthropic Argument (3.4) focuses on the harmfulness of the human species such that it is immoral to continue to bring new members of it into existence. This argument also appears to remain intact if universalism is true. The Misanthropic Argument implies that there is a sense in which Christians who procreate are creating additional harmers. Even if harm is not the only thing that Christians should be concerned with, they ought to care about reducing the amount of harm and suffering in the world as much as possible. The vast amount of harm that is the direct result of human actions would cease if procreation ceased.

Adding considerations about Christian universalism to the mix does not straightforwardly change the strength of the argument. It does not change the amount of harm that the human species perpetrates during their earthly lives. It is a separate question whether the harm could be offset or outweighed by the existence of a good afterlife. I will discuss this below in 6.3.3.

[1] For more on weighing harm and pleasure see Lougheed 2023.

The Hypothetical Consent Argument and Christian Universalism

The Hypothetical Consent Argument (4.3) is such that a person cannot consent to receiving a *pure* benefit. It seems that this argument remains intact even if universalism is true and a good afterlife is guaranteed. This is because a person who is created in a world where universalism is true will still experience harm in their earthly life. This does not change just because they will experience an eternally good afterlife. In fact, creating a person who will experience harms throughout their life in order to get the benefit of a good afterlife appears to be a perfect candidate for a pure benefit. To see this, consider the following modified case:

> *Good Afterlife.* An eccentric spirit knows there is a good heaven and wants to populate it with people. For various reasons, the only way of giving prospective people an afterlife is by creating them on an island that is rife with danger. They will have to toil very hard to provide food for themselves, they will fight amongst each other, 30% of people will get terminal cancers, and rates of other deadly diseases are high. Almost everyone born on the island will experience significant harms in their lifetime. But the only way the spirit can populate a good heaven is by creating persons who must first live on this island. Figuring that everyone who is experiences harm on the island is better off because they will end up in heaven, he proceeds to create people on the island. Many of the inhabitants suffer harms on the island before they die and go to heaven. (Inspired by DeGrazia 2012, 151–152)

This is a case of bestowing a pure benefit since the action is not undertaken in order to avoid a greater harm. No one is harmed by not being brought into existence. The guarantee of a good afterlife does not provide any unique reasons to reject the Hypothetical Consent Argument, and so Christians must grapple with this argument on its own terms.

The Exploitation Argument and Chrisitan Universalism

The Exploitation Argument (4.4) says that babies are exploited in order for parents to get fully formed adults. In the previous chapter, I gestured at a new kind of exploitation argument that relies on the idea of participating in a kind of systemic evil by procreating in a system where some people end up in hell even if your offspring does not (5.4.4). However, there is a different type of exploitation argument available if universalism is true. On this view, it is not babies that are exploited or persons in a problematic system where some people end up in hell. Instead, this style of argument is based on the idea that persons have to exist on earth, enduring all of the harms that come with that, in order to end up in heaven.[2] Procreators are thereby exploiting people in order to populate heaven.

The strength of this argument is probably connected to the degree to which one thinks existing in our world is a harm. Thus, the strength of this argument is connected to how one assesses the evidence offered in the Quality of Life Argument (3.3) and the Misanthropic Argument (3.4). Though some of us may escape the worst harms, it is an empirical fact that most of us will experience at least some of the following: assault, bereavement, betrayal, war, famine, unemployment, homelessness, cancer, heart disease, HIV/Aids, not to mention the vast array of other lesser harms. Though no specific individual will inevitably experience a specific harm before ending up in heaven, the majority of people in heaven will have experienced significant harms in the course of their earthly lives. The people on earth are thereby exploited in order for heaven to be populated. A natural response is that though people must experience harm in their earthly lives, they are not exploited since they end up experiencing the great good of existence of in heaven. I will discuss the degree to which this type of on-balance judgement is successful in 6.3.1.

[2] This argument could be levelled against God in the context of the problem of evil, if indeed God is concerned with populating heaven.

The Argument from the Right to Physical Security and Christian Universalism

It can also be shown that the argument from the Right to Physical Security can be modified to apply to universalism by another modification to Newborn (see 4.6; 5.4.3):

> *Newborn***: Blake has magical powers: With a snap of their finger, a new person appears out of nothing. But Blake's powers work only inside the holodeck, a place where the version of Christianity that is true is always changing due to a randomised computer program. And Blake's creations cannot live outside the holodeck. Unfortunately, the safety protocols are always off, meaning that persons inside the holodeck are vulnerable to whatever version of Christianity is created. The odds of an environment resulting in a world where Christian universalism is true (i.e., where everyone goes to heaven) are at least 80% for any given time someone is on the holodeck. As the holodeck comes to life, Blake snaps their fingers and creates a new person, Anthony, who is in a Christian universalist world and so guaranteed to end up in heaven. If a different version of Christianity had been true, then Anthony may have ended up in traditional hell, hell as deprivation, or annihilated. (Inspired by Hereth and Ferrucci 2021, 20)

Since Blake is responsible for Anthony being brought into a world where Christian universalism is true, though Blake would have partly responsible for Anthony ending up in heaven, Blake also would have been responsible if Anthony was created in a world where a different version of Christianity was true. This means that Blake is responsible for exposing Anthony to a risk of ending up in traditional hell, hell as deprivation, or annihilation. Blake should not expose anyone to a nontrivial (or unknown) risk of ending up in a world where heaven is not guaranteed.

6.2 The Quality of Life Argument for Procreation and Universalism

In this section I develop an argument for pro-natalism, the view that it is permissible to procreate on the assumption that Christian universalism is true. Specifically, I reconstruct the Quality of Life Argument to include the datum of heaven.

The Quality of Life Argument for Procreation

The Quality of Life Argument (3.3) suggests that because humans tend to overrate their abilities, overestimate how well their life is going, Pollyannaism, and because of 'adaptation, accommodation, or habituation', quality of life assessments are unreliable. On hedonistic, desire-fulfillment, and objective list theories of what constitutes a good life, human lives are not very good. In fact, even the very best of human lives are quite bad. This serves to augment the Asymmetry Argument by highlighting the actual harms that people experience in their lives. But it also serves as a standalone argument by suggesting that because humans lives tend to be so bad, they should not be started. However, even supposing one accepted this argument in the original naturalistic context in which it is offered by Benatar, it is reasonable to think that the assessment of the quality of human lives must change if Christian universalism is true. If universalism is true, then every person experiences the great good of an eternally blissful afterlife.

If universalism is true, then regardless of the suffering experienced in a person's earthly life, they are guaranteed to experience a good afterlife for all of eternity. It is hard to understand how to assign a value to blissful afterlife that lasts forever. A diagram with an arrow stretching across a page would not be sufficient to represent it, since the arrow would have to stretch across an infinite number of pages. So, one could agree with Benatar about the Quality of Life Argument on the assumption that naturalism is true and there is no afterlife. But if a good afterlife is guaranteed, then every person's life turns out to be good. The beginning of one's life might be very difficult, but the rest of it, and the part that is the longest—infinitely long—is very good. So, quality of life judgments on the assumption that universalism is true must be extremely positive. Indeed, one might argue that if universalism is true, then every person has an infinitely good life. Call this the Quality of Life Argument for Procreation.

Three Theories of Quality of Life and Universalism

Recall that the three theories of quality of life that Benatar employs in his defense of the original Quality of Life argument are hedonistic theories, desire-fulfilment theories, and objective list theories. With respect to hedonistic theories, the main measure of how well a person's life is going is the extent to which they experience positive mental states, where both the quality and quantity of those states matter. If every person ends up going to heaven, then the degree to which they experience positive mental states is vast. Indeed, the supermajority of their life will consist of positive states. If heaven consists of the strongest quality of such states and is infinitely long, it could well be considered the best possible scenario a person could end up in according to hedonism.

Now, it might be inappropriate to describe the Christian heaven as the best possible version of a kind of pagan hedonistic paradise. Is Christianity compatible with hedonism? Though this is a legitimate question, if there is a heaven, then on virtually any hedonistic metric, lives are going to be going much better than Benatar suggests in his version of the Quality of Life Argument that assumes naturalism is true.

On desire-fulfillment theories, pleasurable mental states matter inasmuch as a person has the desire for such states. The same goes for wanting to avoid negative states. Benatar issues a number of challenges for desire-fulfillment theories, including suggesting that desires can be fulfilled without a corresponding mental state, and that one could be mistaken whether a desire has in fact been fulfilled. He also notes that on the assumption desire fulfillment theories are true, our lives are characterised by unsatisfied desires. However, if there is a blissful afterlife that is guaranteed to everyone, it seems that upon biological death, all (or almost all) of our desires would be fulfilled. Humans would not be mistaken about whether their desires were fulfilled. Desires would also not be temporarily fulfilled in the way that Benatar observes humans tend to experience pleasure sensations.

Again, one might wonder whether a desire-fulfillment theory that measures the quality of life in terms of fulfilled desires is consistent with Christianity. But it is difficult to see how a reasonable set of desires would not be all (or almost all) fulfilled by ending up in heaven. Alternatively, if Christianity is true (and remember the argument assumes Christian universalism is true), then perhaps the correct desire-fulfillment theory will contain those desires most consistent with Christianity.

According to objective list theories, a person's quality of life is measured by the degree to which objective goods obtain in their life. Benatar claims that having pleasurable mental states in addition to the absence of painful mental states will have to be included on any plausible list of objective goods. Though Benatar expresses scepticism about how objective any list of purported goods really are objective, that worry is not relevant to this discussion. If Christian universalism is true, and a list of goods is truly objective, then those goods will necessarily obtain in heaven. An eternally blissful afterlife will, by stipulation, include any and all objective goods. If Christianity is true, it is reasonable to think the list of objective goods will correspond to Christian norms. It is also reasonable to think that obtaining pleasure and an absence of pain will be included in any list of objective goods, even if there are other important items on the list.

Admittedly, there are deeper questions here about the extent to which these theories are compatible with Christianity. And even on the assumption that they are compatible with Christianity, there are still further questions about how to spell out the details of such theories in a way that is consistent with Christianity. However, despite these further questions, the basic point is just that when it comes to measuring the quality of life, no matter how this is done, things improve dramatically if universalism is true. The takeaway is that when on balance judgements about harm and pleasure are permitted (they are not according to the Asymmetry Argument), it appears the scales tip in favour of procreation. If universalism is true, then every life goes extremely well accordingly to any reasonable measurement of the quality of life. Thus, if universalism is true, then procreation is permissible. Contra Benatar, all lives are immensely good if universalism is true.

Some Lives Are Still Not Worth Starting: An Objection to the Quality of Life Argument for Procreation

I have presented what I take to be the strongest possible argument for the permissibility of procreation on the assumption that Christian universalism is true. If universalism is true, then all human lives are immensely good. At first blush, it might seem difficult to see how this argument could be contested. However, the great good of an eternally blissful afterlife does not lend unequivocal support for the permissibility of procreation. To see this, consider the following objection to the scope of the argument.

Recall the case of Dax Cowart that Benatar appeals to in 3.3.1. Part of the point of the example is to suggest that there are some harms that are so bad that any goods to subsequently follow them do not justify or offset those harms. In other words, when an individual looks back on those harms, they determine they were not worth going through even in light of the goods that followed it. Dax ended up living a productive and meaningful life after the pain of the accident and subsequent treatments. Maybe some harms are so awful that no amount of good that occurs after can ever justify them. If something like this is right, then perhaps certain individuals experience so much harm in their lives, even a blissful afterlife that lasts forever does not swamp that harm.

A natural response here is to suggest that Benatar is thinking in terms of finite earthly harms and benefits or pleasure. The goods that Dax experienced after recovering were finite earthy goods. However, if the subsequent good Dax experienced after the accident and treatment was a blissful afterlife, the harms he experienced would have been worth it. A different way of phrasing this response is to consider whether it is permissible to bring Dax into existence with the knowledge of what will happen to him. If there is no afterlife, then probably not (or at least not according to Benatar's original Quality of Life Argument). However, if universalism is true, then it is permissible to bring Dax into existence since after the harms he experiences, he will end up in heaven.

But there are counterexamples to this type of reasoning which show that the intuitions involved here can be contested. It is not obvious that even when a good afterlife exists, that it is permissible to bring any and all types of earthly lives into existence. To understand this, consider the following case:

> *Torture*: A prospective parent knows that if they choose to procreate, their child will be psychologically and physically tortured for the entirety of their 80-year earthly life. There will be doctors and caregivers around, but only to ensure the child is able to grow and develop such that they can continue being tortured. This will continue through adolescence, adulthood, and into old age. The child will not grow up in a loving family environment, they will never attend school or otherwise have the opportunity to develop artistic, musical, or athletic abilities. They will never learn how to read or write. They will never make any friends. The child will never have a family or career of their own. They will grow up in an isolated environment and be kept alive for the sole purpose of being tortured. The prospective parents also know with epistemic certainty that Christian universalism is true and that even though their child will live this unbearably awful earthly life, they will end up in heaven after they endure it.

I submit that it is impermissible to procreate in Torture, especially given that the prospective parent knows what awaits their child once it is born. Indeed, it is morally heinous to procreate in Torture. The earthly life described is simply so horrific, that even if an eternity in the beatific vision awaits them, it is better that they never come into existence. They are not harmed by not being brought into existence. And even if on balance weightings of harm to pleasure are allowed, it does not seem to justify procreating just because the person would get an eternally good afterlife. This is so even if the afterlife is immensely or infinitely valuable. The prior suffering makes it moot.

To further test the strength of the relevant intuition I am driving at in Torture ask yourself the following question: If you were in the womb and offered a choice between ceasing to exist or being born into Torture, would *you* want to be born? I take it that most of us will respond by choosing non-existence. Imagine that it is the *worst possible* existence you could have for 80 years before ending up in heaven. While this is an extreme example, my point is just to demonstrate that universalism simply cannot be used as a trump card to justify procreation in any type of world.

The strength of this objection to the Quality of Life Argument for Procreation depends on what quantity and quality of harms a person would have to experience in their earthly life such that heaven would not justify them coming into existence in the first place. Or alternatively, it depends on what harms a parent would be justified in subjecting a prospective child to who is guaranteed to go to heaven. I do not know how to answer this question. However, I suspect that lives far better than the one describe in Torture, though that still contain much suffering, are *not* justified in starting even if there is a heaven. Finally, notice that if one accepts Benatar's Quality of Life Argument for anti-natalism, then perhaps all or almost all earthly lives contain enough harm such that they are not worth starting, even with the added guarantee of a good afterlife. After all, Benatar argues that most people have lives that are far worse than they realise.

6.3 The Argument from Intrinsic and Final Value and Universalism

In this section, I develop a different argument for the permissibility of procreation that has no parallel anti-natalist argument as located in the concept of intrinsic value. To say that something is intrinsically valuable is to say that the value of a thing is contained within itself. If something is extrinsically valuable, then its value is located outside of it (Korsgaard, 1983, 169–170). If something is instrumentally valuable, it is valuable for the sake of something else. A thing has final value if it is valuable for its own sake (Korsagaard, 1983, 169–170). There is much more that could be said about the nature of intrinsic value, including how to define it more precisely and how such value arises. I think one particularly attractive account says that "something is intrinsically valuable (or good in itself) if and only if it would be valued for its own sake by fully informed, properly functioning valuers" (Davison, 2012, 12; see also Anderson, 1993, 2–3; O'Neill, 1992, 117–18; Rabinowicz & Rønnow-Rasmussen, 2004; Zimmerman, 2001, 105). This means that if humans are intrinsically valuable, then they "would be valued for their own sake by all fully informed, properly functioning valuers, whether they be human, divine, angelic, alien, or whatever" (Davison, 2012, 12). I will not spend time defending this particular account of intrinsic value. Instead, I note that the argument I offer in the next section is likely consistent with a wide variety of different positions on intrinsic value (and the intrinsic value of humans).

The Argument from Intrinsic and Final Value

An argument for the permissibility of pro-creation based on the idea that humans have intrinsic and final value might look something like this:

> *Value Assumption*: All humans are intrinsically and finally valuable. Their value is located within themselves, and they are valuable for their own sake.
> (1) If humans are intrinsically and finally valuable, then creating more humans adds value to the world.
> (2) Humans are intrinsically and finally valuable. (Assumption)
> (3) It is permissible to add (positive) value to the world.
> Therefore,
> (4) It is permissible to procreate.

For the sake of argument, I am assuming that humans are intrinsically and finally valuable. Since I am assuming that Christian universalism is true, this is a safe assumption. I do not expect this to be a place where Christians would object to the argument. The Christian tradition emphasizes the irreducible value of humans. One prominent explanation of this value is based on the idea of *imago dei*, the claim that humans reflect the image of God (see Soskice, 2013). There is much more that could be said on this topic, but doing so would take me too far afield. I am simply going to assume that humans possess intrinsic and final value. If there are ways of interpreting the Christian tradition that minimizes the value or outright denies it, such views are very much in the minority. With this assumption in the background, the rest of the argument is straightforward. Since humans are valuable, then adding humans to the world necessarily adds more value to the world. A world with no humans is less valuable than a world with ten humans, and that world is less valuable than a world with one hundred humans, and so on. Since adding positive value to the world is obviously permissible, it is permissible to procreate.

The Value of Humans and Quality of Life

The Argument from Intrinsic and Final Value does not rely on any particular claims about universalism. However, I offer it in this chapter since in the context of universalism it might be less susceptible to counterexamples. Universalism might be thought to act as a safeguard against claims that creating valuable persons who will suffer in their earthly life is still not justified. But the quality and quantity of harm someone experiences in their earthly life still matters. Consider that as it stands, the value added to the world by creating a person would justify procreation in Torture (6.3.3). But this is absurd. The world would not be better off just in virtue of another human being added to it when they are going to be tortured for their entire earthly life. The quality of a person's earthly life still seems to matter in value assessments even if it is stipulated that people are intrinsically and finally valuable and will eventually end up in heaven.

Additionally, the value of the world and the permissibility of procreation might come apart in cases like Torture. A person might maintain that the child brought into existence is valuable such that their existence on its own adds value to the world in some sense while maintaining that it is still impermissible to bring them into existence. Indeed, the very reason that a person has such value might explain why a prospective parent ought to refrain from bringing someone into existence in Torture. This means that the argument needs an additional qualification:

(5) It is only permissible to create humans if they have a life on earth that is worth living.

But we have learned from Benatar that 'a life worth living' is ambiguous between a life worth starting and a life worth continuing (see 3.2.1). Since this argument is in the context of procreation it needs to be the case that:

(5*) It is only permissible to create humans if they have a life on earth worth starting.

If the Asymmetry Argument is not in view, then this leads right back to questions about the appropriate assessments of the quality of life. I have not issued a definitive verdict on this question. However, even assuming that Benatar's Quality of Life Argument is too strong, I have shown above (6.3.3) that the assumption of universalism does not mean that it is permissible to bring a human into existence in *any* set of circumstances.

A slightly different worry says that it is a mistake to tally human lives in this way. Human lives cannot be added up such that a world with one billion humans is less valuable than a world that is otherwise identical but has ten billion humans. Both worlds are valuable, but neither is more or less valuable than the other. Lurking nearby are questions about Parfit's paradox. If the Argument from Intrinsic and Final Value is sound, it could imply that a world with billions of individuals with lives barely worth starting is better than a world with fewer individuals who all have lives that are intuitively more worth starting because they experience minimal amounts of harm, etc.[3]

An Objection: Mandatory Procreation?

To conclude this section, I offer an objection which intends to show that the Argument from Intrinsic and Final Value proves too much. Suppose someone admits that while some lives are not worth starting, Benatar's

[3] It would take much more discussion to work out in more detail how the distinction between a life worth starting and a life worth continuing might affect the discussion.

suggestions about the quality of typical human lives are far too strong if a good afterlife is guaranteed. Many, if not most, human lives are indeed worth starting because they have on balance very good lives in light of going to heaven. The bliss and joy of heaven makes the vast majority of the typical harms experienced by humans justified. The problem with this line of reasoning is that it is difficult to see why humans are not obligated to create *many more* children. If universalism is true such that all humans brought into existence go to heaven, why not try to create as many humans as possible?

The defender of the Argument from Intrinsic and Final Value might counter that the argument is only intended to show that procreation is permissible, and that nothing they have said implies it is mandatory. But then why is it good to procreate on this view? Why bring children into the world? Is it just selfishness on the part of prospective parents who want the experience of childrearing or help on the farm or care during old age? I think a tempting explanation for procreation, especially in the Christian context, is that a significant reason why it is permissible to procreate is that humans are valuable. But then, this implies endorsing something like the Argument from Intrinsic and Final Value. Again, the response may be to insist that the argument only shows that procreation is permissible, not required.

But I believe the Argument from Intrinsic Value and Final Value needs to be reformulated. From (1) and (2) it does not just follow that it is permissible to procreate, it follows that:

(3*) It is *good* to add (positive) value to the world.
Therefore,
(4*) It is *good* to procreate.
(5*) should also be modified to:
(5**) It is only *good* to create humans if they have a life on earth that is worth starting.

If humans are valuable, then creating more of them is not just permissible, it is good. It is good to add more positive value to the world, even if how to precisely understand this kind of value addition is unclear. This modification explains why procreating is good, especially from a

Christian perspective. Humans are valuable image bearers of the divine. Furthermore, most of the harms that humans experience in a typical earthly life will be offset by the great good of an afterlife. Still, the defender of the Argument might say that just because it is good to φ, it does not follow that there is a requirement to φ. As finite beings with limited time and resources, humans are often forced to choose between different goods. Sometimes these goods compete with each other such that it is impossible to have them at the same time. Having children may well compete with the good of say, a demanding career as a cancer research scientist, where spending long hours in the laboratory will greatly increase the chance of a cure.[4] Further still, perhaps some goods are incommensurate with one another such that one is free to pick from any of them (see Chang, 1998). If having children is like this, then if universalism is true, individuals are permitted to have children provided they have a life worth starting. Having children is good and adds value to the world. But there is no requirement on Christians to have children.

Christians, however, ought to be concerned with adding goodness to the world (perhaps also in addition to ensuring that heaven is populated, though I will not defend this claim here). If Christianity is true, I think that this is a truism. So, the following assumption should be added to the argument:

> *Addition Assumption*: Christians ought to add to the goodness of the world when possible.

But humans do not just have intrinsic and final value such that it is good to bring them into existence but still optional, particularly when doing so competes with other valuable pursuits. The Christian tradition holds that they are the *most* valuable thing in the (material) world. And Christians should be concerned with adding the most positive value to the world, all else being equal. So, the Addition Assumption and Value Assumption should really read:

[4] This assumes that one also wants to be a caring and attentive parent to one's own offspring.

> *Addition Assumption**: Christians ought to add the most goodness to the world when possible.
> *Value Assumption**: All humans are intrinsically and finally valuable. Their value is located within themselves, and they are valuable for their own sake. Humans are the most valuable thing in the material world.

The problem is that this now starts to look as if procreation is not just permissible but is required because bringing new people into the world adds the most value to it. Thus, the Argument from Intrinsic and Final Value should really be:

> (6) If procreation is good and adds the most value to the world, then Christians ought to procreate.
> (7) Procreation is good because humans are the most valuable thing in the (material) world.
> Therefore,
> (8) Christians *ought* to procreate (as much as possible).

This is intended as a *reductio ad absurdum* even though I suspect some Christians might be willing to bite the bullet about the conclusion. One might push back against the Addition Assumption* which is the key to making this objection succeed. However, adding positive value to the world is required by the Christian ethic of self-sacrificial love which means putting the needs of others ahead of your own. If adding more humans to the world adds value to the world, then one ought to procreate as much as possible even if it comes at a great cost to oneself. The quality of earthly life for the children only needs to be balanced against the fact that they are guaranteed to end up in heaven. If humans really are intrinsically and finally valuable and are guaranteed to end up in heaven, then Christians ought to be procreating as much as possible. Indeed, this also means that pursuits that are incompatible with child-bearing and childrearing are impermissible. Or at least this is so on the reasonable assumption that Christians ought to look after their children. If someone admits that Christians ought to procreate but denies they must raise and look after their children, especially if they are pursuing other good things, this equally amounts to a *reductio ad absurdum* of the argument. The right conclusion cannot be that Christians ought to procreate as much as possible but otherwise have no obligations to their offspring.

6.4 A Risk-Based Argument Against Procreation on Universalism

I spent the bulk of Chapter 5 developing risk-based arguments for anti-natalism on the assumption that Chrisitan exclusivism is true. I suggested that if there is a chance one's offspring could end up in hell, where this is understood as eternal conscious torment, it is impermissible to procreate. I underscored the fact that we have a very low risk tolerance with respect to the possibility of harming non-consenting individuals. The reader may have noticed that some of these risk-based arguments could actually be reinterpreted to be apply to the odds of universalism being true. This is troubling for the majority of Christians who think procreation is permissible, since the risk-based arguments on the assumption that exclusivism is true are particularly powerful. Recall the Master Argument from Risk:

> (9) It is impermissible to non-consensually impose a risk of *hell* on others when there is a *significant, or non-trivial, or unknown risk* one will end up in hell, and there is insufficient justification for the imposition of that risk of that harm.
> (10) Bringing a child into existence involves non-consensually imposing a *significant, or non-trivial, or unknown* risk of hell on that child, with insufficient justification for the imposition of that risk of that harm.
> Therefore,
> (11) It is impermissible for the Christian exclusivist to bring children into existence.

This argument can be reformulated to be about the risk that one does *not* end up in heaven and experience eternal bliss:

> (12) It is impermissible to non-consensually impose a risk of *not ending up in heaven* when there is a significant, or non-trivial, or unknown risk one will not end up in heaven, and there is insufficient justification for the imposition of that risk of that harm.

(13) Bringing a child into existence involves non-consensually imposing a *significant, or non-trivial, or unknown* risk of *not ending up in heaven* on the child, with insufficient justification for the imposition of that risk of that harm.
Therefore,
(14) It is impermissible for the Christian universalist to bring children into existence.

Call this 'the Risk-Based Argument for Anti-Natalism from Doubt' (or the 'Argument from Doubt').[5] For those Christian universalists who are not epistemically certain in their universalism, as I suspect most are, this argument holds water. If there is a chance that universalism is false, then there is a chance that not all humans end up going to heaven. This means that by procreating, one necessarily imposes the risk of harming a non-consenting adult. In other words, if there is a chance that humans are not intrinsically valuable, that Christian universalism is false, etc., then procreation is impermissible.

Now, one key difference between the Master Argument from Risk and the Argument from Doubt is that the latter is about the risk of exposing a non-consenting individual to a harm while the former is about the risk of an individual being *deprived* of a benefit. Being deprived of a benefit is bad since there would be someone who already exists and is deprived of it (unlike in the context of the Asymmetry Argument). Still, it seems intuitively obvious that it is worse to be in a position with a risk of ending up in hell as opposed to not ending up in heaven. Thus, if this were all the information one had in assessing these arguments, it could be reasonably objected that they are hardly parallel arguments for Christian anti-natalism, where the former is far stronger.

The way to demonstrate the strength of the Argument from Doubt is to establish the alternatives to heaven. This might seem like an impossible task if one is discussing logically possible alternatives. It is logically possible that when I die, I am reborn as an alien who lives on Jupiter in an alternative universe. These types of possibilities are innumerable. But remember that the context of this book is *Christian* anti-natalism.

[5] Again, see Blackshaw forthcoming for a related point about credences.

Thus, I need only consider the alternatives to heaven that are prominent within the Christian tradition. Doing so will close the gap between these arguments in terms of the degree of exposure to the risk of harm versus the deprivation of a benefit. The following alternatives to heaven are the ones that need to be considered within the Christian tradition[6]:

> Traditional hell: conscious eternal suffering.
> Hell as deprivation: deprived of goods associated with heaven (e.g., God's presence, the beatific vision).
> Purgatory: there is a traditional hell and heaven, though some individuals need to spend time in purgatory to further develop their soul before entering heaven.
> Annihilation: ceasing to exist upon death.

I do not think that these arguments change if there is purgatory. If there is purgatory, there is still a chance of ending up in hell unless one is epistemically certain that universalism is true (i.e., that there is a purgatory that ensures everyone will end up in heaven). The exposure to the relevant set of risks does not change. This leaves traditional hell, hell as deprivation, and annihilationism. If the alternative to ending up in heaven involves a risk of ending up in traditional hell, where this is eternal conscious torment, then the two arguments are indeed parallel. Though not quite as strong, if the alternatives to heaven are hell as deprivation or annihilation, fairly close parallels can still be established. While these alternatives are *comparatively* less bad than traditional hell, they are still quite bad. Hell only as deprivation of the Christian goods associated with heaven is terrible on the assumption that Christianity is true. If Christianity is true, then being in the presence of God's glory forever is the best possible fate. It is bad for anyone to be deprived of this good. Also important is that to be deprived of such an outcome would also make the harms one experiences in their earthly life unjustified. Part of the universalist's strategy is to suggest that the great good of heaven

[6] One might wonder why I have not discussed a bodily resurrection since this is a live option within the Christian tradition. However, I take the concept of heaven I am working with to be consistent with there being a bodily resurrection where there is a new heaven and new earth (perhaps merged into one).

justifies the suffering individuals unavoidably experience in their earthly lives. But if one does not end up in heaven, then this justification is unavailable.

Similar remarks apply to annihilationism. It is comparatively less bad than traditional hell. But it is still quite bad to cease to exist, especially if one had a life worth continuing. Furthermore, if one is annihilated upon biological death, then yet again a person's suffering in their earthly life cannot be justified by appealing to the great good of heaven. In sum, though it is clearly less bad to end up in a hell that is just deprivation or to be annihilated then it is to end up in traditional hell, these results are still harmful. Not only are these outcomes bad in themselves, but they eliminate the possibility of appealing to a blissful afterlife as justification for the harms experienced during one's earthly life. In other words, if Christianity is true and so sets the possible alternatives, the possibility of not going to heaven turns out to be exposure to significant harms, even though some of the possibilities are comparatively less bad than traditional hell.

Recall the appeal to second-hand smoke and driving while intoxicated from Chapter 5. Even if these activities only slightly increase the chance of cancer or of a car accident, people tend to think it is impermissible to impose an increased risk of their occurrence on non-consenting adults. To make these cases more similar to universalism consider the following adjustments to them.

Suppose people knew that being exposed to second-hand smoke either significantly decreased one's chances of getting lung cancer or significantly increased it. Further suppose that it was also known the effects of second-hand smoke are not neutral; it either increases or decreases the chances of lung cancer. In such a scenario, would it be permissible to smoke around a non-consenting person thereby possibly exposing them either to the benefit of being less likely to get lung cancer (which is analogous to heaven) or to the harm of being more likely to end up with cancer (which is analogous to traditional hell, hell as deprivation, and annihilationism)?

As with the risk-based arguments from Chapter 5, I recognise that intuitions may well very about what sort of risk exposure is reasonable to subject someone else to in these types of cases. But I submit that even if the odds of second-hand smoke causing lung cancer were quite low, with the odds of gaining immunity from it quite high, many people would still advise caution. Furthermore, the harm in question in the case of an afterlife is far worse than getting lung cancer. It is eternal conscious torment, eternal deprivation of Gods presence, or non-existence. When a universalist procreates, they might place a high probability on their offspring ending up in heaven but should recognise they are also exposing them to these other terrible outcomes. Since no one is harmed by not being brought into existence, it is better not to expose a potential non-consenting adult to the deprivation of not ending up in heaven, especially when the alternatives are clearly harmful.

Finally, notice that Torture can also be modified to:

> *Torture**: A prospective parent knows that if they choose to procreate, their child will be psychologically and physically tortured for the entirety of their 80-year earthly life. There will be doctors and caregivers around, but only to ensure the child is able to grow and develop such that they can keep being tortured. This will continue through adolescence, adulthood, and into old age. The child will not grow up in a loving family environment, they will never attend school or otherwise have the opportunity to develop artistic, musical, or athletic capabilities. They will never learn how to read or write. They will never have any friends. The child will never have a family or career of their own. They will grow up in an isolated environment and be kept alive for the sole purpose of being tortured. The prospective parents also know that Christian universalism is *probably* true and that even though their child will live this unbearably awful life, they will *probably* end up in heaven after they endure their earthly life. However, there is still a small chance that their child ends up in traditional hell, hell as deprivation, or being annihilated.

Modifying this example to suggest that the parents are not certain that universalism is true, such that after terrible suffering on earth there is no guarantee their child will end up in heaven shows even more clearly that procreating in this scenario is morally heinous. Though one might respond that most typical lives on earth do not contain such monstrous suffering, this has to be balanced against the risk of hell. I submit that *no earthly life is so good that it would be justified creating if one knew in* *advance the person would face traditional hell, hell as deprivation, or being annihilated.*[7] If one does not end up in heaven, it is not permissible to create a person. This does not imply that a life, once started, could not be meaningful even if a person is annihilated, or perhaps even if they end up in hell. So, this does not imply that certain lives are not worth continuing on naturalism or exclusivism. However, it does imply that they should not be started, even though they may be worth continuing. What Torture* demonstrates is that it may be impermissible to procreate if heaven is not guaranteed, even if the risk of ending up in an alternative is relatively low. I do not need to specify the precise risk allowable in order to make this case. I suspect that most universalists will have priors about the chances of ending up in heaven or elsewhere such that any commonsense risk assessment implies that procreation is impermissible.

6.5 Connections to Pascal's Wager

In both Chapter 5 and this chapter, one might have noticed potential connections to Pascal's Wager.[8] I have been intentionally vague about the specific value I assign to heaven, noting that unless one has a credence of 1 that their offspring will end up there, it is impermissible to procreate. However, one common way of interpreting the Wager says that heaven should be assigned a value of infinite utility or as Pascal says, 'an infinity of infinitely happy lives' (Hájek & Jackson, forthcoming). This should have a swamping effect on one's decision-making since ∞ multiplied by any positive probability is ∞. If there is any positive chance that God exists (and there is a blissful afterlife, etc.), then one ought to wager on God.[9] The expected utility of heaven is infinite. Alternatively, the expected utility of hell is $-\infty$.[10] Of course, perhaps only traditional hell

[7] If one insists that this only true of traditional hell, the claim still succeeds if there is a chance of ending up in traditional hell (even if there is also a chance of ending in up in hell as deprivation or being annihilated).

[8] Bruce Blackshaw develops an argument for anti-natalism based on the possibility of hell with explicit reference to Pascal and expected values (forthcoming).

[9] Assigning infinite utility to heaven does not come without problems. To avoid these some instead assign an 'arbitrarily large but finite utility' to heaven (Hájek & Jackson, forthcoming).

[10] It seems that Pascal assigned a finite negative value here (Hájek & Jackson, forthcoming).

should be assigned $-\infty$, where hell as deprivation and annihilationism should be assigned large finite negative values. Still, it is coherent to say that all have a value of $-\infty$ even though traditional hell is the worst.

However, stating more clearly that heaven has a value of infinite utility does not constitute a challenge to any of the risk-based arguments. This is because the Wager takes place in the context of already existing people who now must choose whether to wager on God. I make no comment on whether the Wager succeeds, instead my point is that even if it does, it is not applicable in the context of anti-natalism. In this scenario, the question is whether it is morally permissible to procreate in the first place. This means that the question is whether it is permissible to bring an individual into the world who will face an afterlife with a value of either ∞ or $-\infty$ (or some large finite negative). Once understood in this way, the plausibility of the risk-based arguments remains intact. An individual is not harmed by not being brought into existence and it is better not to face any risk of ending up in an afterlife with a value of $-\infty$. This is so even if there is a chance of ending up in heaven with a value of ∞.

6.6 Conclusion

This chapter explores arguments for anti-natalism in the context of Christian universalism. This is the view that in the final analysis, regardless of the specific mechanics, every person ends up in heaven. I began by claiming that Asymmetry Argument, the Misanthropic Argument, the Exploitation Argument, and the Right to Physical Security Argument are not obviously defeated if Chrisitan universalism is true. Christians need to grapple with these arguments in their own right.

I then examined some positive arguments for pro-natalism on the assumption that universalism is true. I suggested that while quality of life assessments certainly change if a person's life includes heaven, a guarantee of heaven still does not justify starting every life. Some earthly lives are so terrible that even the assurance of heaven at the end of them cannot make them worth starting. I then explored the Argument from Intrinsic and Final Value which suggests that it is permissible to procreate because of the value that humans add to the world. I argued that one problem with this argument is that it proves too much in implying that Christians have an obligation to procreate as much as possible.

To conclude this chapter, I focused on explaining why a risk-based argument similar to the ones I developed in Chapter 5 can still be generated even if universalism is true. If there is a chance that universalism is false, then the universalist exposes their offspring not just to the risk of being deprived of heaven, but to the Christian alternatives of traditional hell, hell as deprivation, and annihilationism. Though the universalist may deem the risk of ending up in one of these alternatives significantly lower than the exclusivist places the odds of ending up in traditional hell, ordinary risk assessments still suggest that even a low risk of the alternatives make procreation impermissible. Unless one is epistemically certain universalism is true, there is a strong risk-based argument to refrain from procreating.

Chapter 5 and this chapter constitute the main defense Christian anti-natalism. I have intentionally not evaluated the arguments for anti-natalism that I deemed compatible with Christianity in their own right. For example, I have not stated whether the Asymmetry Argument or Misanthropic Argument are sound. Instead, what I have argued is that they are compatible with Christianity such that the Christian cannot simply dismiss them for having an offensive conclusion. They must grapple with these arguments. Instead, the main arguments for Christian anti-natalism I developed are risk-based. There are two main takeaways from these arguments: (i) There is a risk of a bad afterlife outcome on both exclusivism and universalism; and (ii) ordinary risk tolerance for imposing the possibility of harm on a non-consenting person are extremely low. These two claims provide extremely strong reasons in favour of anti-natalism, from the Christian perspective.

I have not argued for the claim that Christians should be epistemically cautious about their credence in universalism or exclusivism. In other words, I have not argued that the most rational stance for Christians to take is that there is a possibility of not ending up in heaven, where the alternative outcomes are all bad. It would be a separate project to evaluate the epistemic rationality of the relevant beliefs about salvation and the afterlife. However, I am assuming that any such study would hardly result with the recommendation that people ought to believe in heaven or hell or annihilationism with a credence of 1. It does not require very much philosophical insight to recognise the highly fragile and fallible epistemic situation of humans, including (if not especially) with respect to religious beliefs.[11] It is a different question the extent to which doubts should affect a person's religious practice, if at all. At first blush, at least some degree of doubt is perfectly consistent with living a Christian life devoted to religious practices that seek to honour God. The doubt required to establish these risk-based arguments is very slight, where one only needs to admit there is a small possibility that their stance on the afterlife could be mistaken. This is because there only needs to be a slight risk of not ending up in heaven for the arguments to succeed, given the relevant Christian alternatives.

One might object that my reluctance to assign specific probabilities to various types of afterlives diminishes the plausibility of the argument. But consider that assigning a 0 probability to traditional hell, hell as deprivation, or annihilationism violates *regularity* which says 0 should only be assigned to impossibilities (Hájek and Jackson forthcoming). Additionally, though some argue that you can ignore 'outlandish' or 'farfetched' possibilities, this is hardly uncontroversial (see Kosonen, forthcoming; Monton, 2019; Schwitzgebel, 2017; Wilkinson, 2022). Further, I suspect that as a sociological fact, most Christians, when pushed, would admit that they are not epistemically certain in their beliefs about salvation and the afterlife. I imagine that most would acknowledge that they could possibly be wrong about such beliefs even when they remain very confident about them. Indeed, I suspect that most would not assign 0 or infinitesimally small probability to the afterlives I have mentioned as options in this and the previous chapter.

Even if I am wrong about this, the risk-based arguments for anti-natalism still apply to those Christians who do not assign a 0 or infinitesimally small probability to the alternatives to heaven. So, the arguments still apply to many without the need to convince anyone beforehand that a modicum of doubt about their afterlife beliefs is epistemically appropriate. In sum, though these risk-based arguments will

[11] The existence of widespread religious disagreement alone arguably ought to prompt a healthy degree of epistemic humility. See De Cruz (2019).

admittedly not be convincing to Christians who are genuinely epistemically certain of their beliefs about salvation and the afterlife, such that they do not admit they could *possibly be wrong*, this is hardly a problem for the arguments more generally.

Though I have addressed some specific objections to the arguments developed throughout Chapter 5 and this chapter, in the next chapter I turn to exploring a series of objections to Christian anti-natalism as a position in general. Even supposing a Christian has no specific objections to the risk-based arguments for anti-natalism, the conclusion of anti-natalism is so surprising it should still be expected to raise additional concerns.

7

Objections to Christian Anti-Natalism

7.1 Introduction

Though I have addressed certain objections along the way in Chapters 5 and 6, in this chapter I focus on a series of general objections to Christian anti-natalism. These are not objections to any particular argument for anti-natalism, but instead they are objections to the conclusion of Christian anti-natalism in general. I begin by examining the objections that Christian anti-natalism wrongly assumes there is such a thing as *the* Christian ethic (Sect. 7.2), that it violates procreative freedom (Sect. 7.3), that it wrongly assumes welfarism (Sect. 7.4), and that Christianity is at its core, pro-creationist (Sect. 7.5). I close by addressing a number of less compelling objections such as the claim that anti-natalists have bad characters, that it recommends abortion, and that it misunderstands the nature of Christian marriage (Sect. 7.6). I conclude that none of these objections are decisive, and that Christians must continue to grapple with the arguments for anti-natalism.

K. Lougheed, *Christian Anti-Natalism*, https://doi.org/10.1007/978-3-032-03902-6_7

7.2 There Is No Such Thing as *the* Christian Ethic

One objection to Christian anti-natalism says that the arguments for anti-natalism assume that there is something like an agreed upon interpretation of *the* Christian ethic. There needs to be agreement on what constitutes a Chrisitan ethic in order for there to be agreement about whether the arguments for anti-natalism succeed (or at least remain on the table) within the Christian context. But there is no such thing as *the* Christian ethic within the Christian tradition.

Reply:

This may well be true, but Christians only need to affirm basic normative claims such as *harms (and experiencing harms) are bad*. Any plausible interpretation of the Christian ethic will affirm this and related claims. There does not have to be agreement on what constitutes *the* Chrisitan ethic in order for the arguments for anti-natalism to get off the ground.

7.3 Anti-Natalism Violates Procreative Freedom

At the beginning of this book, I made clear that I am *not* advocating for any anti-natalist policies (Sect. 1.4). I am exploring the best theoretical arguments for the conclusion that Christians should be anti-natalists. Still, the conclusion in question says that Christians should not procreate. One might object that such a conclusion, if accepted, violates the right to procreative freedom. This objection does not rely on any claims about the intrinsic value of humans or about divine commands to procreate. Instead, it relies on the simple claim that humans are entitled to procreative freedom.

Writing in the context of universalism, Blake Hereth suggests that "[a]utonomy is central to procreative ethics, with most supporting a strong *pro tanto* right to reproduction" (2022, 106). This right is typically understood as resting on the more fundamental right to autonomy over one's body.[1] Hereth suggests this argument can be standardised as follows:

The Autonomy Argument

(1) Persons have a strong *pro tanto* right to procreate.
(2) If (1), then procreation is generally permissible.
(3) If anti-natalism is true, then procreation is not generally permissible.

Therefore,

(4) Anti-natalism is false (2022, 107).

If this argument is sound, then anti-natalism is false and it remains permissible for Christians to procreate.

Reply:

Hereth offers an innovative reply to this objection within the context of universalism. Hereth suggests that given the arguments for anti-natalism, it is impermissible to create in one's earthly life. However, Hereth argues that procreative autonomy can be preserved if there is the option to procreate in heaven. On the assumption that no one is harmed by procreation in heaven, then procreation in heaven would be permissible.[2] This would allow an individual to ultimately avoid any harms associated with not procreating on earth (e.g., a violation of their autonomy, missing out on the experience of childrearing, etc.) (Hereth, 2022, 111).[3] Waiting to procreate until one is in heaven is a very marginal harm relative to the fact that heaven is infinitely long and

---

[1] For example, Hereth points to Meijers (2020), Silvers and Francis (2017), and Robertson (1994).

[2] This raises questions regarding the problem of evil. If it is possible to create humans in a place where there is no harm, why would not God create humans in that sort of place to being with? J.L. Mackie suggests a related challenge when he wonders why God could not create libertarian free creates who just never happen to go wrong. See Mackie 1983. For a reply see Plantinga (1989).

[3] Indeed, Hereth observes that whether being unable to conceive a child is a harm is a matter of debate. See Kollar and Loi (2015), Casal and Williams (1995), And Maung (2019), Kukla (2019), Padela and Clayville (2018).

there will be ample (or unlimited) opportunities to procreate.[4] If procreation is possible in heaven, this seems to solve the autonomy challenge.[5] Premise (3) is really about whether it is permissible to procreate in our lives, given the harms that every human will unavoidably face in their life. It does not say anything about procreation in contexts where it is guaranteed that life will be good and there will be no harm associated with procreating for the parents, their offspring, or anyone else. Indeed, Benatar says that his "argument does not apply to those hypothetical cases in which a life contains only good and no bad. About such an existence I [Benatar] say that it is neither a harm nor a benefit and we should be indifferent between such an existence and never existing" (Benatar, 2006, 29).

Notice that Hereth's response only works if (i) universalism is true; and (ii) procreation is possible in heaven. Thus, even if it succeeds as a response to the Autonomy Argument, its scope is necessarily limited. A stronger response is to simply reject (1) or at least suggest that the *pro tanto* right is defeated or outweighed in light of the arguments for anti-natalism. A person's right to autonomy is hardly absolute. One obvious constraint on autonomy is that a person cannot act in ways that harm innocent and non-consenting persons.[6] Furthermore, it is impermissible to exercise one's autonomy if doing so exposes an innocent and non-consenting person to the risk of non-trivial harm. Indeed, one way of reading the risk-based arguments offered in Chapters 5 and 6 demonstrate that no such right to procreative autonomy exists.

This could well be a stalemate, where both sides maintain the other does not appreciate the force of their position. However, I have offered arguments based on the harm and exposure to the risk of harm for anti-natalism, if either exclusivism or universalism is true. The objector who wishes to assert that the right to procreative autonomy defeats the arguments for anti-natalism needs to explain *how* they are defeated. Notice

[4] Hereth also suggests that this would allow humans to fulfill the command to be fruitful and multiply. He suggests this is best interpreted as a command in the context of the garden of Eden (Hereth, 2022, 212). I challenge this interpretation fruitful verses in 2.2.

[5] For arguments for the possibility of procreation in heaven see Worsley (2019) and Hereth (2020).

[6] This is similar to Mill's harm principle. See Mill (1859).

that this involves explaining why the right to procreative autonomy outweighs the right of innocent non-consenting persons to be harmed or exposed to harm. They need to explain what makes procreative autonomy different from other situations where I suspect many would be reluctant to give autonomy that kind of overriding power. Thus, I think it is a very tall order to show how the right to procreative autonomy is a successful objection to anti-natalism. I leave it to proponents of this objection to develop it further.

7.4 Anti-Natalism Assumes Welfarism Which Is Incompatible with Christianity

Welfarism is the view that "all well-being has noninstrumental value, and nothing else has noninstrumental value. This means that all well-being is good independently of its effects, and nothing else is. Though other things can be good or bad, their goodness or badness depends on their contribution to well-being" (Woodard, 2019, 5).[7] Alternatively, welfarism "says that morality is all about individual welfare. It says that facts about the best interests of individuals are, in some sense, the building blocks of morality. Welfarists believe that the whole point of morality is to make individuals better off" (Keller, 2009, 82–83).[8]

If welfarism reduces to human well-being, then clarity is needed about what precisely constitutes well-being. While well-being is often identified as a set of subjective experiences, desires, or a set of objective goods, Simon Keller argues that welfarism can actually remain neutral about well-being:

> [W]e must understand it [i.e., welfarism] simply as the idea that morality is centrally concerned with the best interests of individuals. Welfarism identifies individual welfare, imprecisely, as the central concern of morality, or as the thing that morality is 'all about', but the exact form

[7] Keller points to Sen (1985, 21–22).

[8] Keller notes that many definitions of welfarism are not actually compatible with each other. He demonstrates this by comparing the definitions from Sumner (1996, 184), Moore and Crisp (1996, 598), Kagan (1998, 48) (Keller, 2009, 86).

> of that concern and the sense in which it is central to morality – the more precise sense in which morality is all about welfare – can be specified only in relation to certain independent theoretical commitments, on which different welfarists can disagree. (Keller, 2009, 87)

There are a number of reasons for rejecting welfarism including that the moral good of a person cannot be reduced to welfare, we ought to care about something other than a person's good when evaluating her life morally, and that it might be the case that certain things do not really have interests but that we still have moral obligations to them (Keller, 2009, 90). However, I am not evaluating the merits of welfarism in this section. Instead, I draw attention to it because one might object to Christian anti-natalism by suggesting that it entails welfarism, and that welfarism is incompatible with Christianity. A standardised version of the objection might look like the following:

The Welfarist Argument

(5) If welfarism is incompatible with Christianity, then Christians should reject it.
(6) Welfarism is incompatible with Christianity.

Therefore,

(7) Christians should reject welfarism.
(8) Anti-natalism implies the truth of welfarism.

Therefore,

(9) Christians should reject anti-natalism.

Reply:

In defending (6), the proponent of this argument could claim the following. Welfarism implies that all good or positive value reduces to *human* well-being. But if Christianity is true, this cannot be right because there is good that cannot be reduced to human well-being or perhaps even well-being more generally (if indeed it makes sense to speak of well-being apart from humans). God is good, quite apart from well-being.

If no humans existed, God would still be good. Indeed, some in the Christian tradition would define God as *the* Good, maybe even claiming that God is identical with the Good. To think that goodness reduces to human well-being is mistaken if Christianity is true. If this is the best way of understanding welfarism, then I do not contest that it is incompatible with Christianity.

Rather, the problem with this argument is that (8) is false. I do not believe that anti-natalism implies the truth of welfarism. Alternatively, the truth of anti-natalism does not depend upon the truth of welfarism. Indeed, though I am concerned in this book with the question of human procreation, many of the arguments for anti-natalism apply to any and all sentient beings.[9] To further understand why welfarism is not required for anti-natalism, consider the main normative claim that undergirds most of the arguments is that *harms (and experiencing harms) are bad*. Qualifications could be added such that harms should be avoided, when possible, the risk on them not forced on others, etc. Welfarism cannot be the only normative position that claims as much about harms. Indeed, any intuitively plausible view of normativity will hold that, all else being equal, harms are bad. Anti-natalism does not need to be committed to welfarism.

7.5 Christianity Is Fundamentally Pro-Creationist

Perhaps the most pressing objection to Christian Anti-Natalism focuses on the fact that Christianity is fundamentally and at its core, pro-creation. This can be shown by reflecting on the claim that if God created the world, including humans, then procreation is permissible for God. Since Christians affirm that God created the world, it is necessarily the case that it is permissible for God to bring humans into existence in a world where they will inevitably experience harm. Though the initial act of God's creation involves creating humans in morally perfect conditions (i.e., the garden of Eden), God knew that they would hardly stay

[9] Benatar says the same (see 2006, 2).

perfect.[10] It is safe to say, then, that God is comfortable with procreation in our world, morally speaking.

The Procreation Is Permissible Argument

The kind of reasoning I have in mind can be standardised as follows:

(10) If God creates the world, then it is permissible for humans to procreate.
(11) God creates the world.

Thus,

(12) It is permissible for humans to procreate.

Premise (11) is a truism on Christianity. Not only do Christians hold that God created the world, but many claim that God actively sustains the existence of the world. Premise (10) is where the philosophical action lies.

There is a sense in which this challenge just reduces to the problem of evil (see Blackshaw forthcoming). Indeed, Jason Marsh argues that endorsing the problem of evil entails anti-natalism (2015). If the world is (probably) too evil to have been created by an omniscient, omnipotent, and omnibenevolent God, then why is it permissible for humans to bring other people into existence in such a world? Marsh suggests that instead of embracing anti-natalism, beliefs about the problem of evil need to be revised, along with providing better justifications for procreation than those currently on offer. Thus, one puzzle that emerges when reflecting on (10) is how to consistently be an atheist who endorses the problem of evil while avoiding anti-natalism. Such a position involves embracing the following asymmetry:

[10] This claim assumes a view of divine foreknowledge called Molinism. However, even if Open Theism is true, God still has incredible predictive powers such that God knew the fall was extremely likely. God would also have been aware of the *risk* of the fall and all that it would entail.

> *Evil Asymmetry*: The amount and kinds of suffering we see provide strong, and many will say decisive, evidence against the very idea that our world is the product of divine creation. But this same suffering does not have any bearing on the general morality of human procreation whatever. (Marsh 2015, 15)

So, one atheist response is simply to reject (9) that it is impermissible for God to create people in such a world, and so the existence of our world is evidence that God does not exist. The tension that emerges is for the atheist who wants to avoid anti-natalism. I will address reasons for thinking the Evil Asymmetry is true in the reply below. Instead, the objector may insist that a more plausible principle is:

> *Evil Symmetry*: The amount and kinds of suffering we see provide strong, and many will say decisive, evidence against the very idea that our world is the product of divine creation. This same suffering bears on the general morality of human procreation, suggesting that it is impermissible. (The considerations about evil and harm that apply to human procreation ought to apply to divine creation.)

The Procreation is Permissible Argument seems to me to be the most compelling objection to Christian Anti-natalism. If Evil Symmetry is true, then anti-natalism entails atheism. God is fundamentally a God of creation and the reasons for anti-natalism that should apply to God apparently do not if Christianity is true.

The Anti-Natalism Entails Atheism Argument

This leads to the following problem:

(13) If it is impermissible for humans to procreate, then it is impermissible for God to procreate.
(14) It is impermissible for humans to procreate (anti-natalism).

Thus,

(15) It is impermissible for God to procreate.

Thus,

(16) God does not exist.

The permissibility of human procreation and divine creation go hand in hand. What is permissible or impermissible for the one applies with equal force to the other. This means that Christians cannot, by definition, be anti-natalists. Either anti-natalism and atheism are true or anti-natalism is false and theism is (possibly) true.

Reply:

In order to maintain the possibility of a coherent Christian anti-natalism, it needs to be shown that the moral permissibility of divine creation and human procreation can come apart. One of the more promising ways of doing this involves reflecting on the very different epistemic positions of God and humans. Perhaps divine foreknowledge is what justifies God in creating. God knows the truth about the afterlife, exclusivism, and universalism, etc. God knows whether some harms are so great that there is no way to adequately compensate them. God knows whether any instances of evils are gratuitous. Perhaps in light of this information, God is justified in creating the world. The risk-based arguments do not apply to God because God already knows the results of whatever wager is in view. Since humans are not in the same excellent epistemic position, the risk-based arguments still apply. Unlike God, humans are not sure about the truth regarding the afterlife, exclusivism, and universalism. Now, a tempting response is to suggest that if God created the world, then on the assumption that the considerations of the risk-based arguments hold, universalism must be true. However, such reasoning would only apply to those who are epistemically certain that God exists. For the Christian who admits there is a chance God might not exist, even only a very slight one, then they cannot be certain God created the world. So, for reasons of different epistemic situations, one might say that Christians should not procreate while maintaining that this does not imply God did not create the world.

The Doubt-Based Argument

These considerations lead to the following argument:

(17) If there is a chance that God did not create the world, then Christians should not procreate.
(18) There is a chance that God did not create the world.

Therefore,

(19) Christians should not procreate.

The most obvious drawback to this response is that it seems to admit that anti-natalism indeed entails atheism. It just hinges on the fact that Christians cannot be epistemically certain that atheism is not true. This type of response says that Christians should not procreate if they are not sure that atheism is false. It is unsurprising if the Christian objector finds this to be a rather unpalatable response to the objection.

Another response comes from more carefully analysing the nature of divine foreknowledge. It could be argued that my first reply to this objection assumes Molinism, the view that God knows the truth value of all counterfactuals (see Perszky, 2013). Indeed, Molinism is so strong as to suggest that God knows what humans would freely do in any possible world. Furthermore, God knows which of those possibilities will become actual. Some versions of Open Theism, on the other hand, deny that God knows what future will become actual. There is no fact of the matter about which future will become actual. God's foreknowledge just is awareness of the relevant possibilities. But God does not know which possibility will become actual, even if God has great predictive powers regarding the possibilities of various outcomes. If Open Theism is true, then when God created the world in a morally perfect state, God did not know in advance that it would become fallen. God also did not know just how bad the world would become after it fell. On this view, though it is true that God may have taken a significant risk in creating the world, God did not know the sort of world post-fall humans would inhabit. However, any human considering procreation today know they are bringing their offspring into a fallen world.

Now, more remains to be said here. For example, even on Open Theism, God presumably knew there was a chance the world would become fallen, and that people would experience great harm throughout the course of their earthly lives (see Boyd, 2001). Is this really any different in terms of exposure to risk than what is found in the risk-based arguments for anti-natalism? If the type of exposure to risk is relevantly similar, then appealing to Open Theism cannot justify God in creating the world.

Still other strategies are worthy of exploration. Perhaps responses to the problem of evil can provide reasons why God can procreate but not humans. For instance, given God's great power, maybe God is able to ensure that there are no instances of gratuitous evil or that every evil a person experiences is sufficiently compensated. Of course, it would also have to be shown why humans cannot also appeal to such reasons to justify procreation, and how the reasoning involved does not just reduce to similar considerations to the ones mentioned above. For example, is the only reason humans could not also appeal to responses to the problem of evil that there is a possibility such responses could be wrong?

As stated, I think this is the most pressing argument to any version of Christian Anti-Natalism in general. The Christian God is fundamentally a God of creation. It needs to be shown that human procreation is relevantly morally different from divine creation. If not, then the arguments for anti-natalism I have raised in this book, including the risk-based arguments, turn out to constitute an argument for atheism. The Christian must show where they are flawed or at least make the case that Christianity is more likely to be true than them. For those who cannot see the flaw in either of the arguments for Christian anti-natalism or this objection to it, then they must face this as an unresolved philosophical puzzle that threatens to produce cognitive dissonance.

7.6 Other Objections

In this section, I address a number of other objections to Christian anti-natalism. I believe these all have ready answers and do not require quite as much attention as the four objections I address above.

People with bad characters are Anti-Natalists

One objection to Christian anti-natalism says that anti-natalists tend to have bad characters. A study by Meehan et al. suggests that anti-natalism is associated with "antagonistic narcissism, psychopathic meanness, psychopathic disinhibition, and antagonistic Machiavellianism" (2022, 947).[11] Christians should avoid holding positions that could foster such clearly anti-Christian character traits.

Reply:

This amounts to an *ad hominen* attack against Christian anti-natalism. Even if this were true, it would not show that Christian anti-natalism is false. Instead, it might suggest that Christians who adopt anti-natalism need to be especially vigilant in guarding against developing such traits. This type of response does nothing to show where any of the arguments, including the risk-based arguments, fail. Furthermore, studies of this kind have inherent limitations. Does a study like this show that individuals who already have those dark personality traits are more open to adopting anti-natalism? Or does it show that once it is adopted, anti-natalism will tend to foster those dark personality traits? Perhaps a large longitudinal study could in principle make some progress on these questions, but practically speaking I suspect that doing so would be extremely difficult. The sample sizes of any such study are small. Could there be a selection bias regarding the anti-natalists who are willing to publicly discuss their anti-natalist views? Finally, I am doubtful that any such sample could confidently be thought of as representative of all anti-natalists. And if anti-natalism were widely adopted by Christians, there

[11] See also Schönegger et al. (2023).

is even less reason to think it would be representative, given that it is an extreme minority view in the Christian tradition.

Anti-Natalism Recommends Abortion Which Is Impermissible for Christians

Another objection I suspect that some readers may have had along the way is that anti-natalism recommends abortion which contradicts standard Christian teaching prohibiting abortion. I will not outline why virtually all Christian traditions prohibit abortion, except by noting that many affirm human life begins at conception. I will accept that as stipulated for the sake of argument. If anti-natalism recommends abortion, then Christians cannot adopt it.

Reply:

Though some anti-natalists claim that their arguments extend to abortion, with Benatar as the most obvious example, this need not be the case (2006, 132–160). If human life begins at conception, then the distinction between a life worth continuing and a life worth starting is relevant. On the assumption a zygote or fetus is a human, they may well have a life worth continuing in light of their prospects. There may also be a strong presumption to let them evaluate the quality of their life and decide for themselves whether it is a life worth continuing (a question that most people answer in the affirmative). However, this does not mean they have a life worth starting. And not starting their life would not be killing them or harming them otherwise since non-existent beings cannot be harmed.

What the Christian anti-natalist position can reasonably be accused of recommending in practical terms is abstinence from sexual intercourse or perhaps sterilization. Other contraceptive techniques could be deemed less suitable because there is still a risk of conceiving on them which means the risk-based arguments would probably apply. But nothing here commits the anti-natalist to recommending or permitting abortion.

Anti-Natalism Is Incompatible with Christianity Because God Is Responsible for the Virgin Birth of Jesus

A theologically motivated objection says that Jesus was born of a virgin.[12] God the father made Mary pregnant through a miracle. But this means God is directly responsible for bringing a human into existence. Since God's actions are necessarily morally permissible (if not necessarily ideal), then if God procreates, it must be permissible for humans to procreate.

Reply:

This is a special case that does not generalise. God became incarnate in order to fulfill a very specific mission. On the Christian view, it is good that this happened once humanity needed saving.[13] But if humanity had not required saving, it may not have happened. But more importantly, God becoming incarnate is *not* like a typical instance of procreation. Jesus already exists prior to becoming incarnate and so is not coming into existence during incarnation. This is very different from creating a human who would otherwise not exist. However, if the objector replies that God is still ultimately responsible for creating humans, and this shows procreation is permissible, then see Sect. 7.5.

The Possibility of Procreation Is Fundamental to a Christian Definition of Marriage

Another objection is that procreation is an essential part of any Christian understanding of marriage. Minimally, the possibility of procreation needs to be a part of any Christian marriage. The force of this objection may be felt the most from those Christians within the Roman Catholic tradition. For example, of marriage and procreation, the Catholic Catechism says:

---

12 Depending on the Chrisitan tradition in question, the immaculate conception of Mary can also be added here.

13 This may require rejecting the theological position *felix sulpa*, which says the fall was happy because it led to the great good of incarnation.

> By its very nature the institution of marriage and married love is ordered to the procreation and education of the offspring and it is in them that it finds its crowning glory. Children are the supreme gift of marriage and contribute greatly to the good of the parents themselves. God himself said: "It is not good that man should be alone," and "from the beginning [he] made them male and female"; wishing to associate them in a special way in his own creative work, God blessed man and woman with the words: "Be fruitful and multiply." Hence, true married love and the whole structure of family life which results from it, without diminishment of the other ends of marriage, are directed to disposing the spouses to cooperate valiantly with the love of the Creator and Savior, who through them will increase and enrich his family from day to day. The fruitfulness of conjugal love extends to the fruits of the moral, spiritual, and supernatural life that parents hand on to their children by education. Parents are the principal and first educators of their children. In this sense the fundamental task of marriage and family is to be at the service of life. Spouses to whom God has not granted children can nevertheless have a conjugal life full of meaning, in both human and Christian terms. Their marriage can radiate a fruitfulness of charity, of hospitality, and of sacrifice. (2000, 1652–1654)

Reply:

It is difficult to massage this understanding of marriage as consistent with Christian anti-natalism. It might be that certain branches of Christianity are simply inconsistent with anti-natalism, given their understanding of marriage. However, if a Christian finds themselves compelled by the arguments for anti-natalism, then I submit that they need to reconsider this definition of marriage. This naturally warrants a much larger discussion about the appropriate nature of Christian marriage. However, it is too easy to dismiss the arguments offered in Chapters 5 and 6 as mistaken just because they are incompatible with certain definitions of marriage located within the Christian tradition.

Christians Need to Have Children in Order to Win the Culture Wars

It is possible a Christian might defend means to end reasoning about procreation by appealing to the need to win the culture wars. The idea may be along the lines of needing to have enough Christians in a society to preserve religious freedom such that Christians are able to pursue the lives they see fit to pursue. But the claim may also be significantly stronger. It may instead be that Christians need to procreate to ensure that society and its institutions either keep Christian values or adopt them. Not only is the best kind of society one that fully adopts Christian values, but the best way to protect and foster Christians in their spiritual lives is from within a society that embraces Christianity in the public sphere. In order to do this, Christians need to procreate. Indeed, on some versions of this objection, not only do Christians need to procreate, but they also need to procreate more than other groups in their society.

Reply:

Obviously, this book is not a work of political philosophy, and this objection cuts to the very heart of the relationship between politics and religion. I therefore only make three modest points by way of reply. First, even if Christians have a duty to engage in the culture wars, it is difficult to see how that duty overrides the duty to avoid imposing harms or the risk of harms on non-consenting persons. The culture wars are not more important than the considerations about harm appealed to in most of the arguments for anti-natalism. Second, it is doubtful the main goal of Christians should be to establish a Christian society where this means institutions adopt Christian values. The two greatest commandments are to love God and love your neighbor. The great commission is about making disciples of Christ. It is unclear that these have very much to do with aspiring to political power. This is so even if it is a truism that Christians need to live in societies where there is freedom of religion. Third, it is also not entirely clear how adopting anti-natalism would lose Christians the culture war. If all Christians were to adopt anti-natalism, then there would not be a next generation of Christians to suffer the results of a lost culture war.[14] It is true that there would not be any Christians left to engage with the culture at large and this is admittedly bad even if one is sceptical about the culture wars per se. But again, it is difficult to understand how this type of consideration would be more important than the types of harms pointed to in the arguments for anti-natalism.

7.7 Conclusion

In this chapter, I responded to a series of objections to Christian anti-natalism as a general position. For instance, though it is true that there is no agreed upon interpretation of *the* Christian ethic, any reasonable interpretation of the Christan ethic will be compatible with most of the arguments for anti-natalism, including the risk-based ones. Perhaps the most persistent objection to anti-natalism is that adopting it violates the right to procreative freedom. However, it is doubtful that such considerations override duties not to impose harm or the risk of harm on non-consenting persons. I also suggest that though welfarism is probably inconsistent with Christianity, anti-natalism need not assume it.

The strongest objection to Christian anti-natalism is that the God of Christianity is a creator. There seems to be an intuitive symmetry between the permissibility of divine creation and human procreation. Since God created the world, it is permissible for the Christian to procreate. If the symmetry is maintained, then either anti-natalism needs to be rejected or atheism adopted. The Christian, then, is reasonable to reject anti-natalism since it implies atheism. In response to this objection, I explored a number of potential ways to break the symmetry including that the risk-based arguments may apply to those Christians who believe Christianity could be false, that there are important differences between God's foreknowledge and human knowledge when procreating, and whether certain responses to the problem of evil might

[14] This need not assume that all Christians have parents who are Christians. It just means that whoever becomes a Christian adopts anti-natalism. I suppose if there were many spontaneous converts after all Christians had ceased to exist because of adopting anti-natalism, then the process of there being no Christians in the world could take much longer.

be available to justify divine creation, but not human procreation. At the very least, Christians must grapple with the tension between the pro-creation found in Christianity and the arguments for anti-natalism, particularly the risk-based ones developed in Chapters 5 and 6. In the next chapter, I conclude the book by discussing what might come next for Christian anti-natalism in particular, and Christian procreative ethics in general.

8

Conclusion: The Future of Christian Anti-Natalism

8.1 Introduction

The purpose of this book has been to explore the best possible defense of the claim that Christians ought to be anti-natalists. The result of my study is not a decisive conclusion, though I hope it is admitted that the case for Christian anti-natalism is perhaps much more powerful than may be suspected at first glance. Christians cannot dismiss anti-natalism quickly out of hand as being inconsistent with their religious convictions. In this chapter, after summing up some of the main points from the book in more detail (Sect. 8.2), I suggest that adopting anti-natalism should cause Christians to advocate for easier adoption processes (Sect. 8.3). I conclude by suggesting a number of avenues for future research including teasing out various ethical assumptions and examining the theological implications of anti-natalism (Sect. 8.4).

K. Lougheed, *Christian Anti-Natalism*, https://doi.org/10.1007/978-3-032-03902-6_8

8.2 Where Things Stand

In Part I, after explaining that the purpose of this book was to offer the best possible defense of Chrisitan anti-natalism, without necessarily endorsing it (Chapter 1), I turned to the history of anti-natalism (Chapter 2). Though historical work on anti-natalism is underdeveloped, I showed the idea is hardly new within the history of Christianity. Likewise, it cannot be assumed without much more argument than is typically offered that the Bible is pro-natalist.

In Part II I then summarised the main arguments in the contemporary philosophical literature on anti-natalism starting with the work of David Benatar (Chapter 3). I suggested the Asymmetry Argument, the Quality of Life Argument, and the Misanthropic Argument are all at least *prima facie* consistent with Christianity. After this, I examined other arguments (Chapter 4), concluding that the Risk-Based Argument, Hypothetical Consent Argument and the Exploitation Argument are all *prima facie* consistent with Christianity. However, suggesting such arguments might be consistent with Christianity does not amount to a positive case for Christian anti-natalism. I turned to developing one primarily based on risk in the next part of the book.

In part III, I develop an argument that it is impermissible for Christians who are exclusivists to procreate (Chapter 5). It is immoral to subject a non-consenting individual to the risk of catastrophic harm. This is so even when the risk of such harm is relatively small or unknown. I proceeded to suggest that appealing to universalism will probably not help (Chapter 6). This is because unless a Christian is epistemically certain that universalism is true, they must admit that there is still a chance they are exposing their offspring to serious harm in the afterlife. Furthermore, even if universalism is certain, there are still some earthly lives that are so terrible, the great good of a blissful afterlife still does not justify bringing them into existence. I conclude that these risk-based arguments constitute the most powerful case available for establishing Christian anti-natalism. Christians need to grapple with the risk-based arguments, in addition to the other arguments for anti-natalism that are not obviously inconsistent with Christianity.

The book concludes by examining a series of objections to Christian anti-natalism as a position in general (Chapter 7). I argued that the strongest objection is that if God created the world, then this implies that humans can procreate. I suggested that either the arguments in the book inadvertently imply atheism or there has to be a way of showing how divine creation and human procreation are not relevantly morally symmetrical.

My main aim in this book has been relatively modest. My claim is just that the arguments for anti-natalism cannot be ignored or dismissed quickly as obviously inconsistent with Christianity. I believe that the risk-based arguments I develop are particularly worthy of consideration. Alvin Plantinga once remarked something along the lines of that the 'arguments for God's existence are as good as any philosophical arguments'. The same, I think, could probably be said about the philosophical arguments for anti-natalism if the position being defended was not so offensive. Anti-natalism cuts across the grain of almost every society that exists or has existed. I hope that this book will play a part in anti-natalism ceasing to be viewed as beyond the pale for reasonable discussion in Christian circles.[1] To reiterate yet again, none of this means that I am an anti-natalist nor does it mean that the arguments I developed in this book are decisive.

8.3 Alternative Ways of Living

One of my secondary motivations for writing this book is to prompt Christians to be more reflective on their procreative choices. Indeed, if Christians did this because of arguments for anti-natalism, I suspect that two byproducts might emerge. First, Christians could become more open to alternative ways of family-making, particularly adoption. Benatar recognises the difficultly of adoption when he writes that:

[1] Some of the arguments I discuss obviously apply to other theistic traditions.

> Adoption is an arduous process with considerable costs of an emotional and financial kind. Most of those who currently endure this process are those who cannot produce their own genetic child, although there are some people who elect to adopt in spite of their own fertility. For both fertile and infertile adopters, non-procreative rearing comes at the cost of the adoption process (Benatar, 2006, 97)

There is little doubt that adoption is a very difficult process in many jurisdictions. However, if more people wanted to pursue adoption, pressure would mount to make the process easier. If Christians were as committed to adoption as having their own genetic children, there would be strong motivation for them to advocate for better adoption processes. Better adoption processes could also have the byproduct of convincing women who would otherwise have an abortion to carry the fetus to term and give it up for adoption. For Christians who are opposed to abortion, this byproduct would be to the good.

Second, when Christians view marriage and having genetic offspring as the best way to live, they risk inadvertently ostracising couples who cannot have biological children and single people who, for whatever reason, never marry and have children. Chrisitan churches and communities are supposed to be welcoming and if Christians slowed down to reflect on their procreative choices, it might lead to better inclusion for those Christians that do not have children.

Commandments to love your neighbor as yourself or to love your enemies suggest something rather radical. If Christians are supposed to treat relative strangers in this way, it is difficult to distinguish in what ways a biological child would or should constitute a special relationship that generates unique duties and obligations. Indeed, that consequentialism denies the moral relevance of special relationships is sometimes considered an objection to it. However, the reasoning for this on the Christian view is entirely different. It is a truism that having children greatly reduces a parent's time and energy.[2] Parents of young children are frequently sleep deprived and run ragged attempting to meet the constant demands of a needy infant or toddler. When it comes to parents

[2] This assumes that a parent is at least trying to minimally fulfill their parental duties.

of biological children, all of this time and energy could be devoted to Christian charity for *people who already exist.* Remember that no one is harmed if a couple does not have biological children. The great resources that Christian parents devote to biological children could be devoted elsewhere if anti-natalism was adopted. Obviously, these claims about the implications of a Chrisitan ethic are hardly uncontroversial. But it is interesting to consider that taking anti-natalism seriously might actually turn out to better promote Christian self-sacrificial love and charity than typical family-making. These topics deserve more attention from scholars working in practical theology.

8.4 The Future of Christian Anti-Natalism

In many ways, I have only offered a bird's eye view of the case for Christian anti-natalism in this book. Many topics need to be examined in much more detail in order to discover the plausibility of Christian anti-natalism.

As I stated in Chapter 2, I am not a Biblical scholar or historian of Christianity. It would be helpful for Biblical scholars to pick up some of the threads and explore the 'fruitful' verses, etc. The history of anti-natalism in general, and within Christianity in particular, remains rather shrouded. Interested historians should take up the task of better tracing anti-natalism movements throughout human civilization.

Turning more to philosophical questions, I have made very minimal claims about the type of morality that is entailed by Christianity. For example, I have only said that it is necessary to think harm should be avoided. It would be interesting to survey the most prominent interpretations of Chrisitan ethics in order to determine whether any of the arguments for anti-natalism get weaker or stronger on different accounts. I have also said very little about different conceptions of harm. Though I think the arguments are going to be compatible with multiple views of harm, it would be helpful for this to be demonstrated more conclusively.

What are the practical implications of Christian anti-natalism, if it was widely adopted? There remain many practical topics that I only briefly gestured at above (Sect. 8.3). What is the appropriate definition and function of Christian marriage if anti-natalism is true? If anti-natalism is true, how should Christian institutions change? What should church look like? I also said very little about formal theories in eschatology. But for those Christians who believe in literal 'end times', etc., what, if anything, changes if Christian anti-natalism is true? Indeed, many theological positions would probably have to be rethought in light of the truth of Christian anti-natalism.

8.5 Conclusion

The arguments for anti-natalism are stronger than is supposed by many Christians. If Christians want to maintain their typically pro-natalist stance, they need to grapple with these arguments, especially the risk-based ones. Despite the fact that I have only set out to develop the best possible case for Christian anti-natalism, not defend it as true, it will come as little surprise to me if this book has made certain readers angry.[3] I hope that such readers will engage this work maturely and respond to the arguments I have developed in peer-reviewed academic venues.[4]

[3] Though I suppose I might be surprised if an angry reader has read to the end.

[4] I would also hope that such readers would pause and reflect on *why* this work has made them angry.

Abbreviations

AE. (1999). *Augustine Through the Ages: An Encyclopedia.* ed. Allan Fitzgerald, Grand Rapids, MI: Eerdmans.

ANF. *Ante-Nicene Fathers.* (1867–1873) 10 vols. ed. Alexander Roberts and James Donaldson. Edinburgh: T&T Clark.

LW. Luther, Martin. (2007) *Luther's Works.* 55 vols. ed. Jaroslav Pelikan. Logos edition. St. Louis: Concordia, 1955.

NPNF. *Nicene and Post-Nicene Fathers.* 28 vols. eds. Philip Schaff and Henry Wace. New York: Scribner, 1886–1900

K. Lougheed, *Christian Anti-Natalism*, https://doi.org/10.1007/978-3-032-03902-6

Bibliography

Adams, M. M. (2000). *Horrendous evils and the goodness of God*. Cornell University Press.

Agar, N. (2010). *Humanity's end: Why we should reject radical enhancement*. MIT University Press.

Agar, N. (2013). *Truly human enhancement: A philosophical defense of limits*. MIT University Press.

Akerma, K. (2020). Kurnig and his neo-nihilism: The first modern antinatalist. In K. Lachmanová (Ed.), *History of antinatalism: How philosophy has challenged the question of procreation* (pp. 125–145). Independently published.

Akerma, K. (2021). *Antinatalism: A handbook*. Epubli.

Akerma, K. (2024). On Matti Häyry's "exit duty generator." *Cambridge Quarterly of Healthcare Ethics, 33*(2), 232–237.

Anderson, A. (1988). Oncomouse released. *Nature, 336*(6197), 300.

Anderson, E. (1993). *Value in ethics and economics*. Harvard University Press.

Andrews, F. M., & Withey, S. B. (1978). *Social indicators of well-being: Americans' perspectives of life quality*. Plenum Press.

Appleby, M. C., Mench, J. A., & Hughes, B. O. (2004). *Poultry behaviour and welfare*. CABI Publishing.

K. Lougheed, *Christian Anti-Natalism*, https://doi.org/10.1007/978-3-032-03902-6

Arnett, J. (2008). The neglected 95%: Why American psychology needs to become less American. *American Psychologist, 63*(7), 602–614.
Asch, S. E. (1956). Studies of independence and conformity: I. A minority of one against a unanimous majority. *Psychological Monographs, 70*(9), 1–70.
Barth, K. (1999). *Church dogmatics Vol. 3 The doctrine of creation* (G. T. Thomson, Trans., ed. G. W. Bromiley). T&T Clark.
Basinger, D. (2020). Religious diversity (pluralism). In E. N. Zalta (Ed.), *The Stanford Encyclopedia of Philosophy* (Winter 2020 ed.).
Bawulski, S. (2013). Do hell and exclusivism make procreation morally impermissible? A reply to Kenneth Himma. *Faith and Philosophy, 30*(3), 330–344.
Bayne, T. (2010). In defence of genethical parity. In D. Archard & D. Benatar (Eds.), *Procreation and parenthood: The ethics of bearing and rearing children* (pp. 31–56). Oxford University Press.
Belshaw, C. (2012). A new argument for anti-natalism. *South African Journal of Philosophy, 31*(1), 117–127.
Belshaw, C. (2015). Immortality, memory and imagination. *Journal of Ethics, 19*(3–4), 323–348.
Benatar, D. (1997). Why it is better never to come into existence. *American Philosophical Quarterly, 34*(3), 345–355.
Benatar, D. (1999). The unbearable lightness of bringing into being. *Journal of Applied Philosophy, 16*(2), 173–180.
Benatar, D. (2000). The wrong of wrongful life. *American Philosophical Quarterly, 37*(2), 175–183.
Benatar, D. (2001). To be or not to have been? *International Journal of Applied Philosophy, 15*(2), 255–266.
Benatar, D. (2006). *Better never to have Been: The arm of coming into existence.* Oxford University Press.
Benatar, D. (2011). The owl and the ostrich: A reply to Sami Pihlström on ethical unthinkabilities and philosophical seriousness. *Metaphilosophy, 42*(5), 605–616.
Benatar, D. (2012). Every conceivable harm: A further defence of anti-natalism. *South African Journal of Philosophy, 31*(1), 128–164.
Benatar, D. (2013). Suicide: A qualified defense. In J. S. Taylor (Ed.), *The ethics and metaphysics of death: New essays* (pp. 222–244). Oxford University Press.
Benatar, D. (2015). The misanthropic argument for anti-natalism. In S. Hannan, S. Brennan, & R. Vernon (Eds.), *Permissible progeny? The morality of procreation and parenting* (pp. 34–58). Oxford University Press.

Benatar, D. (2017). *The human predicament: A candid guide to life's biggest questions*. Oxford University Press.
Benatar, D. (2019). Not "not 'better never to have been'": A reply to Christine Overall. *Philosophia, 47*(2), 353–367.
Benatar, D. (2022). Misconceived: Why these further criticisms of anti-natalism fail. *The Journal of Value Inquiry, 56*(1), 119–151.
Bengtson, V. L., Putney, N. M., & Harris, S. (2013). *Families and faith*. Oxford University Press.
Blackshaw, B. P. (forthcoming). Applying Pascal's Wager to Procreation. *Sophia*.
Blenkinsopp, J. (1997). The family in first temple Israel. In L. Perdue, J. Blenkinsopp, J. J. Collins, & C. Meyers (Eds.), *Families in Ancient Israel* (pp. 8–103). Westminster John Knox.
Blum, D. (1995). *The monkey wars*. Oxford University Press.
Bond, R., & Smith, P. B. (1996). Culture and conformity: A meta-analysis of studies using Asch's (1952b, 1956) line judgment task. *Psychological Bulletin, 119*(1), 111–137.
Boonin, D. (2012). Better to be. *South African Journal of Philosophy, 31*(1), 10–25.
Boothby, N. G., & Knudsen, K. M. (2000). Children of the gun. *Scientific American, 282*(6), 60–65.
Bortolotti, L., & Nagasawa, Y. (2009). Immortality without boredom. *Ratio, 22*(3), 261–277.
Bradley, B. (2010). Benatar and the logic of betterness. *Journal of Ethics and Social Philosophy*, 1–5.
Boyd, G. (2001). *Satan and the problem of evil*. InterVarsity Press.
Bratton, S. P. (1992). *Six billion and more: Human population regulation and Christian ethics*. Westminster John Knox Press.
Braithwaite, V. (2010). *Do fish feel pain?* Oxford University Press.
Brill, S. (2012). Sick and healthy: Benatar on the logic of value. *South African Journal of Philosophy, 31*(1), 38–54.
Brown, C. (2011). Better never to have been believed: Benatar on the harm of existence. *Economics and Philosophy, 27*, 45–52.
Breetvelt, I. S., & Van Dam, F. S. A. M. (1991). Underreporting by cancer patients: The case of response shift. *Social Science and Medicine, 32*(9), 981–987.
Brogan, A. (1919). The fundamental value universal. *The Journal of Philosophy, Psychology and Scientific Methods, 16*, 96–104.
Brooke, A. T. (2012). How best to prevent future persons from suffering: A reply to Benatar. *South African Journal of Philosophy, 31*(1), 79–93.

Buenting, J. (Ed.). (2010). *The problem of hell: A philosophical anthology* (pp. 178–198). Routledge.
Cabrera, J. (1996). *Critica de la moral afirmativa: Una reflexión sobre nacimiento, muerte y valor de la vida.* Gedisa.
Cabrera, J. (2011). *A ética e suas negações, não nascer, suicídio e pequenos assassinatos* (2nd ed.). Rocco. (Original work published 1989).
Casal, P., & Andrew, W. (1995). Rights, equality and procreation. *Analyse & Kritik, 17*(1), 93–116.
Catholic Church. (2000). *Catechism of the Catholic church* (2nd ed.). Liberia Editrice Vaticana.
Chang, R. (1998). *Incommensurability, incomparability, and practical reason.* Harvard University Press.
Chisholm, R., & Sosa, E. (1966). On the logic of "intrinsically better." *American PhilosophicalQuarterly, 3*, 244–249.
Clarke, A., & Haraway, D. (Eds.). (2018). *Making kin not population.* Prickly Paradigm Press.
Coates, K. (2014). *Anti-natalism: Rejectionist philosophy from Buddhism to Benatar*. First Edition Design Publishing.
Cohen, J. (1989). *'Be fertile and increase, fill the earth and master it': The ancient and medieval career of a biblical text*. Cornell University Press.
Conly, S. (2016). *One child: Do we have a right to have more?* Oxford University Press.
Coole, D. (2018). *Should we control world population?* Polity Press.
Cottingam, J. (2005). *The spiritual dimension: Religion, philosophy and human value*. Cambridge University Press.
Cottinghamm, J. (2016). Meaningfulness, eternity, and theism. In J. Seachris & S. Goetz (Eds.), *God and meaning* (pp. 123–136). Bloomsbury Academic.
Craig, W. L. (2013). The absurdity of life without God. In J. Seachris (Ed.), *Exploring the meaning of life: An anthology and guide* (pp. 153–172). Wiley-Blackwell.
Crenshaw, B. R. (2018). *Be fruitful and multiply? Christian theism and the ethics of procreation* (Master's thesis). Denver Seminary.
Crist, E. (2019). *Abundant earth: Toward an ecological civilization.* University of Chicago Press.
Davison, S. A. (2012). *On the intrinsic value of everything*. Continuum.
De Cruz, H. (2019). *Religious disagreement*. Cambridge University Press.
Dever, W. G. (2003). *Who were the early israelites and where did they come from?* Eerdmans.

DeGrazia, D. (2010). Is it wrong to impose the harms of human life? A reply to Benatar. *Theoretical Medicine and Bioethics, 331*, 317–331.

de Giraud, T. (2021). *The childfree christ: Antinatalism in early Christianity*. AFNIL.

de Grey, A., & Rae, M. (2007). *Ending aging: The rejuvenation breakthroughs that could reverse human aging in our lifetime*. Martin's Press.

Doriani, D. M. (1993). Birth dearth or bring on the babies? Biblical perspectives on family planning. *Journal of Biblical Counseling, 12*(1), 24–35.

Doyle, D. J. (2012). Cryonic life extension: Scientific possibility or stupid pipe dream? *Ethics in Biology, Engineering and Medicine, 3*(1–3), 9–28.

Fagley, R. (1960). *The population explosion and christian responsibility*. Oxford University Press.

Feinberg, J. (1992). Wrongful life and the counterfactual element in harming. *Freedom and Fulfilment* (pp. 3–36). Princeton.

Feinberg, J. (1984). *The moral limits of the criminal law Volume 1: Harm to others*. Oxford University Press.

Fischer, J. M. (1994). Why immortality is not so bad. *International Journal of Philosophical Studies, 2*(2), 257–270.

Gerber, D. E. (1999). Greek elegiac poetry: From the seventh to the fifth centuries B.C. Harvard University Press.

Gerstenberger, E. (2001). *Psalms part 2 and lamentations*. Eerdmans.

Greenwald, A. G. (1980). The totalitarian ego: Fabrication and revision of personal history. *American Psychologist, 35*(7), 603–618.

Glannon, W. (2002). Extending the human life span. *Journal of Medicine and Philosophy, 27*(3), 339–354.

Goetz, S. S. (2012). *The purpose of life: A theistic perspective*. Continuum.

Gould, D. C. (2021). Future minds and a new challenge to anti-natalism. *Bioethics, 35*, 793–800.

Gurney, M. E., et al. (1994). Motor neuron degeneration in mice that express a human Cu, Zn Superoxide dismutase mutation. *Science, 264*(5166), 1772–1775.

Hájek, A., & Jackson, E. (forthcoming). Pascalian expectations and explorations. In R. Ariew & Y. Avnur (Eds.), *The Blackwell companion to Pascal*. Wiley-Blackwell.

Hare, C. (2007). Voices from another world: Must we respect the interests of people who do not, and will never, exist? *Ethics, 117*(3), 498–523.

Hallich, O. (2022). David Benatar's argument from asymmetry: A qualified defence. *The Journal of Value Inquiry, 56*, 5–19.

Hamilton, V. P. (1990). *The book of genesis: Chapters 1–17*. Eerdmans.

Harman, E. (2009). Critical study of David Benatar. *Better never to have been: The harm of coming into existence. Nous, 43*(4), 776–785.
Harrison, G. (2012). Antinatalism, asymmetry, and an ethic of prima facie duties. *South African Journal of Philosophy, 31*(1), 94–103.
Hartley, J. E. (2000). *Genesis.* Hendrickson.
Hauskeller, M. (2022). Anti-natalism, pollyannaism, and asymmetry: A defence of cheery optimism. *The Journal of Value Inquiry, 56*(1), 21–35.
Häyry, M. (1994). *Liberal utilitarianism and applied ethics.* Routledge.
Häyry, M. (2004). A rational cure for pre-reproductive stress syndrome. *Journal of Medical Ethics, 30*(4), 377–378.
Häyry, M. (2023). Procreative generosity: Why we should not have children. *Philosophies, 8*(5), 96.
Häyry, M. (2024). Exit duty generator. *Cambridge Quarterly of Healthcare Ethics, 33*(2), 217–231.
Häyry, M., & Sukenick, A. (2024). *Antinatalism, extinction, and the end of procreative self-corruption.* Cambridge University Press.
Hedberg, T. (2020). *The environmental impact of population: The ethics of procreation.* Routledge.
Hereth, B. (2020). The shape of trans afterlife justice. In M. Panchuk & M. C. Rea (Eds.), *Voices from the edge: Centering marginalized perspectives in analytic theology.* Oxford University Press.
Hereth, B. (2022). Heavenly Procreation. *Faith and Philosophy, 39*(1), 100–123.
Hereth, B., & Ferrucci, A. (2021). Here's not looking at you, kid: A new defence of anti-natalism. *South African Journal of Philosophy, 40*(1), 14–33.
Himma, K. (2010). Birth as a grave misfortune: The traditional doctrine of hell and Christian salvific exclusivism. In J. Buenting (Ed.), *The problem of hell: A philosophical anthology* (pp. 179–198). Routledge.
Himma, K. (2016). The ethics subjecting a child to the risk of eternal torment: A reply to Shawn Bawulski. *Faith and Philosophy, 33*(1), 94–108.
Hannan, S., Brennan, S., & Vernon, R. (2015). *Permissible progeny? The morality of procreation and parenting.* Oxford University Press.
Hunter, A. G. (1999). *Psalms.* Routledge.
Jonas, H. (2001). *The gnostic religion: The message of the Alien God and the beginnings of Christianity.* Beacon Press.
Juengst, E. T., et al. (2003). Biogerontology, 'anti-aging medicine', and the challenges of human enhancement. *Hastings Center Report, 33*(4), 21–30.
Kagan, S. (1998). *Normative ethics.* Westview Press.
Kahneman, D. (2013). *Thinking, fast & slow.* Random House of Canada.

Kavka, G. S. (1982). The paradox of future individuals. *Philosophy and Public Affairs, 11*(2), 93–112.

Kaposy, C. (2009). Coming into existence: The good, the bad, and the indifferent. *Human Studies, 32*, 101–108.

Keller, C. (1995). A Christian response to the population apocalypse. In H. Coward (Eds.), *Population, consumption and the environment: Religious and secular responses* (pp. 109–121). State University of New York Press.

Keller, S. (2009). Welfarism. *Philosophy Compass, 4*(1), 82–95.

Kidner, D. (1975). *Psalms 73–150*. Inter-Varsity Press.

Kikawada, I., & Quinn, A. (1985). *Before Abraham was: The unity of genesis 1–11*. Abingdon Press.

Kilmer, A. (1972). The mesopotamian concept of overpopulation and its solution as reflected in the mythology. *Orientalia, 41*(2), 160–177.

Kosonen, P. (forthcoming). Probability discounting and money pumps. *Philosophy and Phenomenological Research*.

Kollar, E., & Loi, M. (2015). Prenatal equality of opportunity. *Journal of Applied Philosophy, 32*(1), 35–49.

Kraus, H.-J. (1989). *Psalms 60–150: A commentary* (H. C. Oswald, Trans.). Augsburg.

Kristof, N. (2009, January 3). If this isn't slavery, what is? *The New York Times*.

Lochmanová, K. (2020). *History of antinatalism: How philosophy has challenged the question of procreation*. Independently Published.

Kukla, R. (2019). Infertility, epistemic risk, and disease definitions. *Synthese, 196*, 4409–4428.

Lachmanová, K. (Ed.). (2020). *History of antinatalism: How philosophy has challenged the question of procreation*. Independently published.

Lachmanová, K. (2020). History of antinatalism: From modern age to present. In K. Lachmanová (Ed.), *History of antinatalism: How philosophy has challenged the question of procreation*. Independently published.

Lachmanová, K. (2020b). Protohistory of antinatalism: Antiquity and the middle ages. In K. Lachmanová (Ed.), *History of antinatalism: How philosophy has challenged the question of procreation* (pp. 37–52). Independently Published.

Le Couteur, D. G., & Barzilai, N. (2022). New horizons in life extension, healthspan extension and exceptional longevity. *Age and Ageing, 51*(8), 1–6.

Lewis, C. S. (1945). *The great divorce*. Geoffrey Bles.

Lougheed, K. (2020a). *The axiological status of theism and other worldviews*. Palgrave Macmillan.

Lougheed, K. (2020b). *The epistemic benefits of disagreement*. Springer.

Lougheed, K., & Hendricks P. (2021). Conceptions of faith. In D. Pritchard (Ed.), *Oxford bibliographies online*. Oxford University Press.
Lougheed, K. (2022a). *Ubuntu and western monotheism: An axiological investigation*. Routledge.
Lougheed, K. (2022b). *African communitarianism and the misanthropic argument for anti-natalism*. Palgrave Macmillan.
Lougheed, K. (2023). Future minds are not a challenge to anti-natalism: A reply to Gould. *Bioethics, 37*(2), 208–213.
Lougheed, K. (forthcoming). Benatar and Metz on cosmic meaning and anti-natalism. *The Journal of Value Inquiry*.
Mackie, J. L. (1983). *The miracle of theism*. Oxford University Press.
Magnusson, E. (2019). How to reject Benatar's asymmetry argument. *Bioethics, 33*(6), 674–683.
Magnusson, E. (2022). On risk-based arguments for anti-natalism. *The Journal of Value Inquiry, 56*, 101–117.
Matlin, M., & Stang, D. (1978). *The pollyanna principle: Selectivity in language, memory, and thought* (pp. 160–166). Schenkman Publishing Company.
Mangina, J. L. (2004). Bearing fruit: Conception, children and the family. In S. Hauerwas & S. Wells (Eds.), *The Blackwell companion to Christian ethics* (pp. 246–265). Blackwell Publishing.
Mays, J. L. (1994). *Psalms*. Westminster John Knox Press.
Marsh, J. (2015). Procreative ethics and the problem of evil. In S. Hannan, S. Brennan, & R. Vernon (Eds.), *Permissible progeny? The morality of procreation and parenting* (pp. 65–86). Oxford University Press.
Maung, H. H. (2019). Is infertility a disease and does it matter? *Bioethics, 33*(1), 43–53.
McFadden, R. D., & Macropoulos, A. (2008, November 29). Wal-Mart employee trampled to death. *The New York Times*.
McGregor, R., & Sullivan-Bissett, E. (2012). Better no longer to be. *South African Journal of Philosophy, 31*(1), 55–68.
McLean, B. (2015). What's so good about non-existence? *Journal of Value Inquiry, 49*, 81–94.
McMahan, J. (1998). Wrongful life: Paradoxes in the morality of causing people to exist. In J. Coleman & C. Morris (Eds.), *Rational commitment and social justice: Essays for Gregory Kavka* (pp. 208–248). Cambridge University Press.
Mill, J. S. (2003). *On liberty* (Original work published 1859). Yale University Press.

Meehan, M. K., Zeigler-Hill, V., & Shackelford, T. K. (2022). Dark personality traits and anti-natalist beliefs: The mediating roles of primal world beliefs. *Philosophical Psychology, 37*(4), 947–969.

Meijers, T. (2020). The value in procreation: A pro-tanto case for a limited and conditional right to procreate. *The Journal of Value Inquiry, 54*, 627–647.

Metzinger, T. (2017). Benevolent artificial anti-natalism (BAAN). *Edge.*

Metz, T. (2011). Are lives worth creating? *Philosophical Papers, 40*(2), 233–255.

Metz, T. (2012). *Meaning in life: An analytic study* (p. 135). Oxford University Press.

Metz, T. (2021). The meaning of life. In E. N. Zalta (Ed.), *The Stanford encyclopedia of philosophy.*

Metz, T. (2022). Does the lack of cosmic meaning make our lives bad? *The Journal of Value Inquiry, 56*, 37–50.

Monton, B. (2019). How to avoid maximizing expected utility. *Philosophers' Imprint, 19*, 1–25.

Moore, A., & Crisp, R. (1996). Welfarism in moral theory. *Australasian Journal of Philosophy, 74*, 598–613.

Milgram, S. (1974). *Obedience to authority*. Harper & Row.

Moran, W. (1971). Atrahasis: The Babylonian story of the flood. *Biblica, 52*(1), 51–61.

Narveson, J. (1967). Utilitarianism and new generations. *Mind, 76*(301), 62–72.

Oberdiek, J. (2017). *Imposing risk: A normative framework.* Oxford University Press.

O'Neill, J. (1992). The varieties of intrinsic value. *The Monist, 75*(2), 119–137. Reprinted in H. Geirsson & M. Losonsky (Eds.), *Beginning metaphysics* (pp. 105–122). Blackwell Publishers, 1998.

Overall, C. (2012). *Why have children?* MIT Press.

Packer, J. (2011). Better never to have been? The unseen implications. *Philosophia, 39*(2), 225–235.

Padela, A. I., & Clayville, K. (2018). Treating infertility with transplantation: Theological views on whether infertility is a disease. *The American Journal of Bioethics, 18*(7), 40–42.

Parfit, D. (1984). *Reasons and persons*. Clarendon Press.

Perszyk, K. (2013). Recent work on molinism. *Philosophy Compass, 8*(8), 755–770.

Pihlström, S. (2009). Ethical unthinkabilities and philosophical seriousness. *Metaphilosophy, 40*(5), 656–670.

Plantinga, A. (1989). *God, freedom, and evil*. Eerdmans.

Plutarch. (1928). *Moralia* (Vol. 2, F. C. Babbitt, Trans.). Harvard University Press.

Poledníková, M. (2020). Solovyov and his godmanhood: Antinatalism in Russia. In K. Lachmanová (Ed.), *History of antinatalism: How philosophy has challenged the question of procreation* (pp. 111–124). Independently Published.

Poulson, A. (2006). *An examination of the ethics of contraception with reference to recent protestant and Roman Catholic thought*. PhD thesis. King's College London.

Polgreen, L. (2008, October 28). Court rules Niger failed by allowing girl's slavery. *The New York Times*.

Provan, C. (1989). *The bible and birth control*. Zimmer Press.

Rabinowicz, W., & Rønnow-Rasmussen, T. (1999). A distinction in value: Intrinsic and for its ownsake. *Proceedings of the Aristotelian Society, 100*(1), 33–51.

Rieder, T. (2016). *Toward a small family ethic: How overpopulation and climate change are affecting the morality of procreation.* Springer.

Robertson, J. A. (1994). *Children of choice: Freedom and the new reproductive technologies.* Princeton University Press.

Schönegger, P., Maier, M., & Bastos, R. V. S. (2023). Introduction and validation of the short antinatalism scale (S-ANS). *Psychological Test Adaptation and Development, 4*(1).

Schwitzgebel, E. (2017). 1% skepticism. *Noûs, 51*(2), 271–290.

Schwitzgebel, E. (2024). Belief. In E. N. Zalta & U. Nodelmann (Eds.), *The Stanford encyclopedia of philosophy* (Spring 2024 ed.).

Sen, A. (1985). *Commodities and capabilities.* Elsevier Science Publishers.

Silvers, A., & Francis, L. P. (2017). Reproduction as a civil right. In L. P. Francis (Ed.), *Handbook on reproductive ethics* (pp. 182–205). Oxford University Press.

Shevelow, K. (2008). *For the love of animals: The rise of the animal protection movement*. Henry Holt and Company.

Smart, R. N. (1958). Negative utilitarianism. *Mind, 67*(268), 542–543.

Smart, R. (1996). Subject selection bias in psychological research. *Canadian Psychology, 7*(2), 115–121.

Smyth, N. (2020). What is the question to which anti-natalism is the answer? *Ethical Theory and Moral Practice, 23*(1), 71–87.

Smyth, N. (2022). Nothing personal: On the limits of the impersonal temperament in ethics. *The Journal of Value Inquiry, 56*(1), 67–83.

Smuts, A. (2014). To be or never to have been: Anti-natalism and a life worth living". *Ethical Theory and Moral Practice, 17*(4), 711–729.

Soskice, J. M. (2012). Human dignity and the image of God. In C. McCrudden (Ed.), *Understanding human dignity* (pp. 229–243). Oxford University Press.

Spurrett, D. (2011). Hooray for babies. *South African Journal of Philosophy, 30*(2), 197–206.

Stagg, N. (forthcoming). Be fruitful, but do not multiply. *Faith and Philosophy.*

Sullivan-Bissett, E. (2022). Better to return whence we came. *The Journal of Value Inquiry, 56*, 85–100.

Sumner, L. W. (1996). *Welfare, happiness, and ethics.* Oxford University Press.

Svoboda, F. (2020). Aristotle and his Eudemus. In K. Lachmanová (Ed.), *History of antinatalism: How philosophy has challenged the question of procreation* (pp. 53–67). Independently Published.

Taylor, S. E. (1989). *Positive illusions: Creative self-deception and the healthy mind.* Basic Books.

Taylor, S., & Brown, J. (1988). Illusion and well-being: A social psychological perspective on mental health. *Psychological Bulletin, 103*(2), 193–210.

Taylor, K., Gordon, N., Langley, G., & Higgins, W. (2008). Estimates for worldwide laboratory animal use in 2005. *Alternatives to Laboratory Animals, 36*(3), 327–342.

Thomson, J. (1986). Imposing risks. In J. Thomson (Ed.), *Rights, restitution, and risk* (Vol. 128, pp. 285–311). Harvard University Press.

Torres, P. (2020). Can anti-natalists oppose human extinction? The harm-benefit asymmetry, person-uploading, and human enhancement. *South African Journal of Philosophy, 39*(3), 229–245.

Van Leeuwen, R. (2001). 'Be fruitful and multiply': Is this a command, or a blessing? *Christianity Today, 45*(14), 58–61.

von Wright, G. H. (1963). *The logic of preference.* Edinburgh University Press.

Walls, J. L. (1992). *Hell: The logic of damnation.* University of Notre Dame Press.

Weinberg, R. (2012). Existence: Who needs it? The non-identity problem and merely possible people. *Bioethics, 27*(9), 471–484.

Weinberg, R. (2021). Ultimate meaning: We don't have it, we can't get it, and we should be very, very sad. *Journal of Controversial Ideas, 1*(1), 1–22.

Weinstein, N. D. (1980). Unrealistic optimism about future life events. *Journal of Personality and Social Psychology, 39*(5), 806–820.

Weinstein, N. D. (1984). Why it won't happen to me: Perceptions of risk factors and susceptibility. *Health Psychology, 3*(5), 431–457.

Westermann, C. (1983). *Genesis 1–11* (J. J. Scullion, Trans.). SPCK.

Williams, B. (1973). The Makropulos case: Reflections on the tedium of immortality. In B. Williams (Ed.), *Problems of the self* (pp. 82–100). Cambridge University Press.

Wilkinson, H. (2022). In defense of fanaticism. *Ethics, 132*(2), 445–477.

Woodard, C. (2019). *Taking utilitarianism seriously*. Oxford University Press.

Worsley, D. (2019). Limbo, hiddenness, and the beatific vision (and procreation, for some, in the life to come). In B. Hereth & K. Timpe (Eds.), *The lost sheep in philosophy of religion* (pp. 347–364).

Yarbrough, O. L. (1986). *Not like the gentiles: Marriage rules in the letters of Paul*. Scholars Press.

Yoshizawa, F. (2021). A dilemma for Benatar's asymmetry argument. *Ethical Theory and Moral PraCtice, 24*, 529–544.

Zimbardo, P. (2007). *The Lucifer effect: Understanding how good people turn evil*. Random House.

Zimmerman, M. J. (2001). *The nature of intrinsic value.* Rowman & Littlefield Publishers.

Index

K. Lougheed, *Christian Anti-Natalism*, https://doi.org/10.1007/978-3-032-03902-6

V

W

Y

Z

FSC
www.fsc.org
MIX
Papier | Fördert
gute Waldnutzung
FSC® C083411

Kolibri 360 GmbH
Ferdinand-Jühlke-Straße 7
99095 Erfurt, Deutschland
produktsicherheit@kolibri360.de